Robert M. Garrels
Fred T. Mackenzie
Cynthia Hunt

# chemical
# cycles
# and the
# global environment

## assessing human influences

WILLIAM KAUFMANN, INC.
One First Street, Los Altos, California 94022

ACKNOWLEDGEMENTS

We are grateful to numerous students and colleagues
for criticisms and suggestions, many of which have been
incorporated in this text.

We thank Jean Kartchner, Leslie Yukuno, Bruce Mitchell
and Cheril Cheverton for the illustrations.  The devil in
Fig. 43 is courtesy of John T. Stark.

ISBN 0-913232-29-7

# PREFACE

Numerous natural substances are important to the
chemistry and biology of the earth--to the earth's
"metabolism."  There is a natural circulation of these
materials through the earth's atmosphere, to the land
and thence through soils into streams, and from streams
and rivers to the oceans.  In the oceans some of the
material sinks and becomes part of the sediments of the
ocean floor, and some of it is returned to the atmosphere.

This circulation of materials is known as the
exogenic cycle.  It has continued for millions of years
at a roughly constant pace.  Over the centuries, mankind's
population and influences on the environment, and hence
on the exogenic cycle, have been growing.

Everyone is aware of the kinds of local environmental
problems caused by many human activities:  pollution from
automobiles, factories, agriculture.  But few are acquainted
with the "everyday geochemistry" that provides a way for
us to assess the ways in which natural environmental
processes and chemical cycles respond to the polluting
effects that are a perhaps inevitable consequence of growing
human populations.

This book grew out of an undergraduate course that
we developed to provide perspective on the innumerable
environmental issues that are raised, chiefly by local
problems.  Our idea was to examine the natural circulation
of materials and then to try to assess man's impact
on these cycles.  We wanted to see whether or not particular
pollutants are having irreversible or reversible effects
on the environment.  A major objective was to discover

those activities that may have irreversible effects and those whose effects, if controlled, may be transitory.

The book reflects the results of our efforts to fulfill these objectives and to provide an introduction to the chemistry of the natural environment. It has been developed and modified over a period of four years as a result of experiences in our own courses at Northwestern University and the University of Hawaii. It has also been influenced by helpful suggestions from teachers at other universities who have used earlier versions of our syllabus.

The present version is organized as follows: After several chapters on the general circulation of materials at the earth's surface and the characteristics of various environments, global cycles of particular substances that are important in the chemistry and biology of the earth are presented in detail. An attempt is made to assess the degree of interference with each cycle resulting from human activities and to determine whether the effects of human additions are short or long-term, trivial or important.

In most examples cited, the earth's surface environment is subdivided into four great reservoirs: the land surface, the atmosphere, the oceans, and the sedimentary rocks. The sizes of these reservoirs and the annual rates of transfer of selected substances among them are estimated. We believe that local and regional pollution problems must be fitted into the global framework. Many polluting substances are high local concentrations, resulting from human activities, of elements or compounds that have been migrating through the major earth reservoirs for hundreds of millions of years.

These substances have natural sources and sinks, and there are feed-back systems that regulate the earth's natural metabolism and that must be considered in investigations of local situations.

We believe this book can be used as a text for environmental courses in almost any science or engineering department. It is also designed to be sufficiently self-contained to be assigned as collateral reading or for independent study by the layman interested in human influences on the world ecosystem. The material has been taught successfully to students ranging from college freshmen with essentially no science or mathematics background to juniors and seniors majoring in chemistry, biology, engineering, earth sciences, physics, or oceanography. The first eleven chapters are appropriate for anyone who has had high school chemistry and algebra. The final chapter may pose some difficulties for students with these minimum requirements but not for students who have an interest in and aptitude for any of the sciences. We regard Chapter 12 as optional. The book is complete without it, but it has been successful in showing students the directions of present-day research and in giving them a feeling for the validity of the data we have used and the kinds of information that are urgently needed to solve many of the problems posed.

Most of the chapters are followed by study questions. Some of the questions contain information not included in the text. Answers to questions are given at the end of the book, rather than with the questions, because students found it tempting to read the answers before thinking about the questions if the two were juxtaposed.

The answers often are our interpretations; we have commonly found student disagreement with them, and they have therefore been useful in stimulating discussion.

In Appendix I, information is presented on the compositions of rocks and minerals; Appendix II contains some convenient conversion factors.

The selected bibliography provides suggested additional reading and includes some of our sources of data. It seemed impractical for student use to document all sources of data used in development of each cycle. The data on which the various cycles are based are being modified and added to daily.

One reason for the book's present format is to permit frequent updating. We are grateful to the publisher for his cooperation in making an experimental edition available. We invite readers to send us their constructive criticisms which might be incorporated into future editions.

Evanston, Illinois
June, 1975

Robert M. Garrels
Fred T. Mackenzie
Cynthia A. Hunt

TABLE OF CONTENTS

Page

Chapter 1    GLOBAL FRAMEWORK                                         1
             Study Questions                                         14

Chapter 2    RESIDENCE TIME                                          16
             Introduction                                           16
             Multiple Fluxes                                        18
             Reservoirs Within Reservoirs                           19
             Residence Time and Pollution                           19
             The Earth System                                       20

Chapter 3    THE ATMOSPHERE                                          21
             Composition, Structure and Circulation                 21
             Pollution and the Atmosphere                           25
             Study Questions                                        30

Chapter 4    THE OCEANS                                              31
             Composition, Structure and Circulation                 31
             Air-Sea Interaction                                    41
             Restricted Marginal Basins                             42
             Food Chain and Element Enrichment                      44
             Study Questions                                        48

Chapter 5    THE GLOBAL CYCLE OF WATER AND OTHER MATERIALS -
             THE EXOGENIC CYCLE                                      50
             The Water Cycle                                        50
                 Lakes                                              54
             The Material (Exogenic) Cycle                          54
                 Introduction                                       54
                 The Steady-State System                            57
                 Feed-back Mechanisms                               64
                 Summary                                            66
             Natural Production of Some Gases                       66
             Study Questions                                        71

Chapter 6    THE CARBON CYCLE                                        73
             Study Questions                                        79

Chapter 7    THE SULFUR CYCLE                                        81
             Global Cycle                                           81
                 Summary                                            85
             Stable Isotopes as an Aid in Pollutional Studies:
             Sulfur as an Example                                   85
             Sulfur Cycling During Geologic Time                    87
                 Sulfur Isotopes at Salt Lake City                  89
                 Summary                                            91
             Study Questions                                        92

Chapter 8    THE NITROGEN CYCLE                                    95
             The Atmosphere                                        95
             Toxicity                                              98
             Land, Sea and Rocks                                   98
             Oxygen, Phosphorus and Nitrogen                      102
             Study Questions                                      103

Chapter 9    THE PHOSPHORUS CYCLE
             Study Questions                                      110

Chapter 10   TRACE ELEMENTS                                       112
             Some General Comments                                112
             Effect of Trace Elements on Man                      115
             Mercury, Lead and Manganese as Examples of
             Trace Element Behavior                               119
                     Mercury                                      119
                     Toxic Concentrations                         124
                     Uses of Mercury                              126
                     Conclusions                                  126
                     Lead                                         127
                     Cycling                                      127
                     Toxicology                                   129
                     Summary                                      130
                     Manganese                                    131
             Study Questions                                      134

Chapter 11   SYNTHETIC ORGANICS, PETROLEUM AND PARTICULATES       136
             Synthetic Organics                                   136
                     DDT Toxicology and Use                       136
                     DDT Global Model                             140
             Petroleum                                            142
                     Behavior in Seawater                         144
                     Effect on Biosphere                          144
             Particulates                                         144
             Study Questions                                      147

Chapter 12   QUANTITATIVE MODELING OF NATURAL CHEMICAL CYCLES     148
             Introduction                                         148
             Mathematical Considerations                          151
                     Steady-State Models                          152
             Examples of Modeling                                 155
                     Mercury                                      155
                     Phosphorus                                   157
                     A Simplified Phosphorus System               160
                     Perturbation of the System                   162
             Perturbation of More Complex Models                  166
             Lead Metabolism in Humans - A Three Reservoir Model  173
             Summary                                              176
             Study Questions                                      178

APPENDIX I    Rocks and Minerals                    179

APPENDIX II   Glossary and Definitions              181

SELECTED BIBLIOGRAPHY                               183

ANSWERS TO STUDY QUESTIONS                          187

## LIST OF FIGURES

Page

Fig. 1    Factors involved in the flow of materials                                  2

Fig. 2    Examples of production and consumption trends for                           4
          raw materials

Fig. 3A   Economic growth rates of nations                                           5
     3B   Graph illustrating rough correlation between
          national per capita GNP and energy consumption

Fig. 4    Carbon dioxide concentration in the atmosphere                             8

Fig. 5    U.S. energy consumption 1850 to 2000                                       11

Fig. 6    Energy requirements for various foods                                      12

Fig. 7    Pressure-height relations in the atmosphere                                23

Fig. 8    Temperature-height relations in the atmosphere                             23

Fig. 9    Vertical distribution of some atmospheric components                       24

Fig. 10   Concentration of dust in the atmosphere                                    24

Fig. 11   Plan view of atmospheric circulation                                       27

Fig. 12   Schematic vertical circulation of the atmosphere                           27

Fig. 13   Wind velocities as a function of latitude, altitude
          and Northern Hemisphere season                                             28

Fig. 14   Distribution of land and sea                                               32

Fig. 15   Temperature relations in ocean and atmosphere                              32

Fig. 16   Typical vertical distribution of nutrients in the
          ocean                                                                      37

Fig. 17   Typical vertical distribution of silica in the oceans                      37

Fig. 18   Global circulation of oceanic surface currents                             39

Fig. 19   Schematic circulation of oceans                                            40

Fig. 20   Block diagram of Atlantic                                                  40

Fig. 21   Latitude variation of precipitation and evaporation
          and gross relation to wind belts                                           43

Fig. 22    Estuarine circulation                                          43

Fig. 23    Classical lagoonal circulation                               45

Fig. 24    Schematic diagram of lagoonal circulation                    45

Fig. 25    Global water cycle                                           52

Fig. 26    Soil water migration                                        52

Fig. 27    Schematic cross section of a continent                      56

Fig. 28    Schematic steady system suggesting average relations
           over the past few hundreds of millions of years             62

Fig. 29    The exogenic cycle                                          65

Fig. 30    The carbon cycle                                            74

Fig. 31    The sulfur cycle                                            82

Fig. 32    The nitrogen cycle                                          94

Fig. 33    Schematic N cycle                                          100

Fig. 34    The phosphorus cycle                                       105

Fig. 35    Volatility of trace elements to the atmosphere            114

Fig. 36    Correlation between trace element concentrations
           in streams and permissible concentrations in water
           supplies                                                  117

Fig. 37    Model of pre-man cycle of mercury                         120

Fig. 38    Model of present-day cycle of mercury                     120

Fig. 39    Prediction of future changes of mercury in surface
           layer of oceans                                           123

Fig. 40    Cycling of lead                                           128

Fig. 41    Pre-man cycle of manganese                                132

Fig. 42    Present-day cycle of manganese                            132

Fig. 43    Simplified DDT cycle                                      141

Fig. 44    DDT concentration in troposphere and oceanic
           organic mixed layer                                       143

Fig. 45    Schematic diagram of a three-box cycling model            145

Fig. 46    Schematic diagram of a three-box, steady state
global cycling model    153

Fig. 47    Global cycle of phosphorus    159

Fig. 48    A.  Oceanic cycle of phosphorus    161
               B.  Formalization of oceanic cycle of phosphorus    161

Fig. 49    Steady state in seawater-phosphorus system    165

Fig. 50    Doomsday scenario    167

Fig. 51    P reservoir changes owing to an exponentially increasing rate of P mining    167

Fig. 52    Lead metabolism in the normal human    174

Fig. 53    Lead metabolism in human after 20x increase in
lead assimilated    177

LIST OF TABLES

| | | page |
|---|---|---|
| Table 1 | General environmental factors | 9 |
| 2 | Energy content of world supply of fossil fuel | 11 |
| 3 | Composition of the atmosphere | 26 |
| 4 | Major dissolved species in seawater | 33 |
| 5 | Composition of streams and the ocean | 33 |
| 6 | Concentrations and concentration factors of some trace elements | 46 |
| 7 | The hydrosphere | 50 |
| 8 | Solar energy budget | 51 |
| 9 | Average composition of earth's crust, sedimentary rocks and suspended load of streams | 56 |
| 10 | Agents of material transport to oceans | 58 |
| 11 | Natural cycles of the major elements | 59 |
| 12 | Reservoir sizes and residence times of the elements | 61 |
| 13 | Gas production to atmosphere | 69 |
| 14 | Emission rates of dominant pollutant types and sources for 1966 and 1968 | 70 |
| 15 | Properties of atmospheric sulfur species | 83 |
| 16 | Comparison of mining production of some metals, metal emission rates to atmosphere owing to man's activities, worldwide atmospheric rainout and total stream load | 111 |
| 17 | Estimates of mercury concentrations in the environment | 122 |
| 18 | Physiological effects of DDT residues and PCB's on organisms in laboratory tests | 137 |
| 19 | DDT and PCB residues in oceanic birds and fishes | 139 |
| 20 | DDT residue concentrations in the environment | 141 |
| 21 | Annual fluxes of petroleum and its by-products | 143 |
| 22 | Fluxes of material brought into the exogenic cycle by man and natural processes | 145 |
| 23 | Phosphorus contents of geochemical reservoirs and inter-reservoir fluxes | 168 |

Chapter 1


GLOBAL FRAMEWORK


Analogy commonly is made between the earth's surface system and a giant chemical-engineering factory; material circulation in the natural system is driven by energy derived from the sun and by decay of radioactive elements in the earth's interior. In another sense, the earth has a natural metabolism; materials have circulated about the earth's surface for millions of years. Weathering and erosion of rocks moves materials in and out of the atmosphere, to the oceans via streams, from the atmosphere to the biota and back again, and to the continents by uplift. Each element follows a path through the natural system determined by its physical-chemical properties; each element has its natural chemical cycle. Man's activities have contributed materials to these cycles at an increasing rate; some of these materials are the same chemical species that have circulated for millions of years; others are synthetic compounds, foreign to the natural environment.

As background for discussion of man's additions to and the consequent effects on natural chemical cycles in subsequent chapters, we consider briefly some of the reasons for growing concern with environmental degradation and interference with these cycles.

Pollution arises from fluxes of materials such as metals, fuels and foodstuffs through the environment. Figure 1 is a schematic diagram portraying material pathways and factors involved in the flow of materials. Although constructed for a specific environment (urban), the diagram illustrates the complex interrelationships between factors involved in the flow of materials and environmental quality. A complete diagram would include flows of materials through the agricultural and rural environments and the interrelations of these with the urban environment.

Consumption of most basic materials such as metals, plastics, quarry stone, and so forth, in the U.S. has been rising steadily; for example, consumption of iron rose 30% between 1950 and 1971. Figure 2 illustrates consumption and production trends for some raw materials. Increased consumption is accompanied

2

FIGURE 1

Factors Involved in the Flow of
Materials.

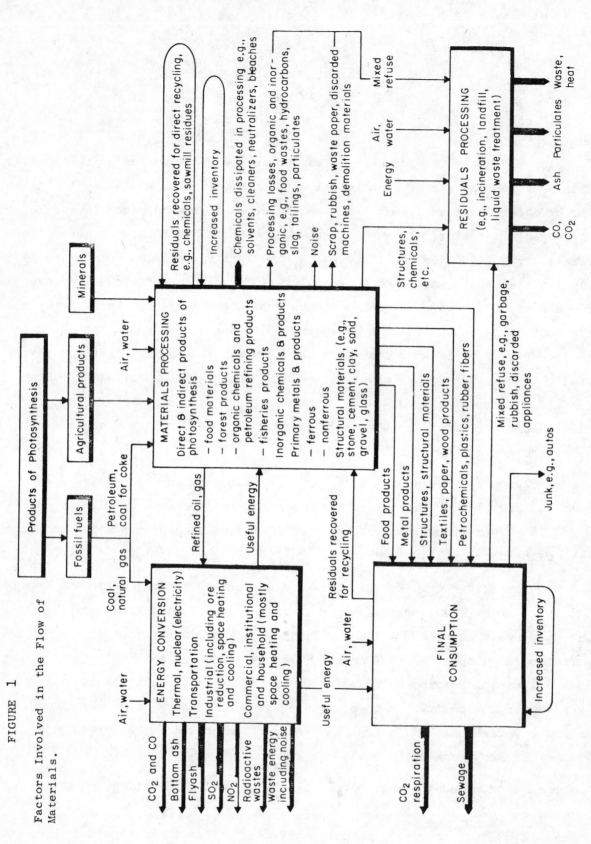

Source: Robert U. Ayres and Allen V. Kneese, "Pollution and Environmental Quality," Quality of the Urban Environment, Harvey S. Perloff, Ed., Resources for the Future, Inc., Washington, D.C., p. 37.

by increased use of fossil fuels and minerals used in energy conversion and materials processing. The residual materials (wastes) of consumption and materials processing include those discarded directly to the physical environment ·(witness the mountains of 'junked' automobiles), and others that may be incinerated, used in landfill projects, or treated in waste treatment plants. Whatever the case, energy conversion, materials and residuals processing, and consumption result in wastes being added to the physical environment as particulate and gaseous emissions, radioactive wastes and chemicals, sewage, and a variety of other solids and liquids. These wastes may eventually reach the ocean via the atmosphere or streams, or in some instances (e.g., petroleum) may be added directly to the oceans. In Figure 1 these wastes, commonly termed pollutants, are represented by the heavy, black arrows. The production and flow of agricultural products also produce wastes that enter the physical environment; e.g., application of DDT to crops results in escape of DDT to the atmosphere and its dispersion throughout the environment; use of nitrogen- and phosphorus-containing fertilizers results in release of these elements to stream systems. It is these types of wastes and their effect on the global environment with which we will be most concerned.

In general, a nation's wastes are proportional to its level of economic activity, or as an approximation of this activity, gross national product (GNP, expressed in the U.S. as total value in dollars of products produced). Economic growth rates of several nations are shown in Figure 3A. Notice that economic growth of the more industrialized nations has progressed, at least until recently, at a faster rate than for less industrialized nations. In other words, the gap between rich and poor has been widening.

A rough correlation exists between per capita GNP (dollars per capita) and consumption of energy, or BTU or kilocalories per year per capita (Fig. 3B). High energy consumption has been a prerequisite for high output of goods and services. Between 1962 and 1968, U.S. GNP/capita and energy consumption both increased 50%, accompanied by a commensurate increase in emissions. In contrast, population growth was less than 2% over this same period, and recently appears to have leveled out.

The rise in GNP/capita and energy consumption is well illustrated by the situation arising from the burning of fossil fuels. Carbon dioxide, as well as other materials, is released to the atmosphere when coal, oil and gas are burned.

4

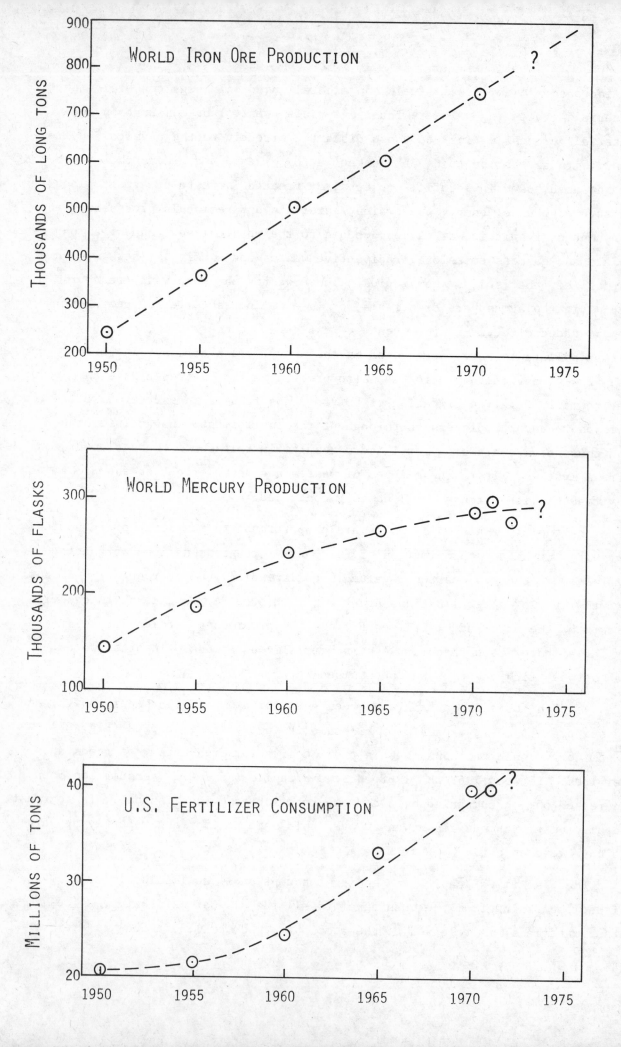

Figure 2. Examples of production and consumption trends for raw materials. (Data from Statistical Abstract of the U.S., 1973; Mineral Yearbook, v.1, 1974).

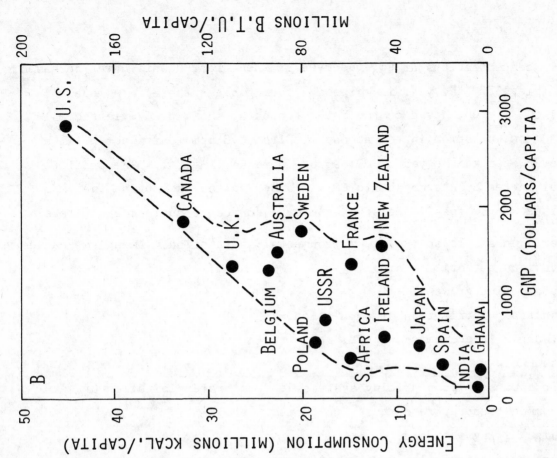

Figure 3B. Graph illustrating rough correlation between national per capita GNP and energy consumption--most nations fall within envelope defined by dashed lines.

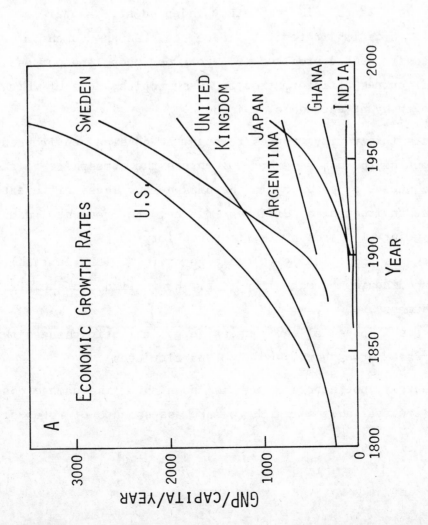

Figure 3A. Economic growth rates of nations (from Kuznets, S., 1971. Economic Growth of Nations, Harvard Univ. Press, Cambridge, Mass.)

Atmospheric concentrations of $CO_2$ observed at Mauna Loa, Hawaii have increased steadily since 1958 (the year in which monitoring began there).  A model of the global $CO_2$ system, based on the Mauna Loa data, shows a rise in the calculated $CO_2$ concentration in the atmosphere from 293 ppm (parts per million) in 1860 to 380 ppm in the year 2000, a 30% increase (Fig. 4).  The predicted exponential increase in atmospheric $CO_2$ is based on the assumption that man will continue to use fossil fuels as an energy source at an increasing rate.

There is no single source of environmental degradation.  Among the causes cited by various authorities are:

1.  population growth,
2.  spatial distribution of population,
3.  changes in economic output,
4.  technology,
5.  economics -- poor pricing systems lead to overexploitation of common property resources (fuel, ores, etc.),
6.  high living standard,
7.  use of fossil fuels -- coal, oil, gas.

None of these factors is completely independent;  all are complexly inter-related, and each is implicated to some degree.  An assessment of each one as a source of environmental problems is beyond our scope and purpose.  There are many good sources of information on these topics, some of which are listed in the bibliography at the end of this book.

Pollutants take various routes through the physical environment -- particles settle out of the atmosphere;  $CO_2$ enters the atmosphere as gas, is dissolved in the ocean, or taken up in the biosphere.  These different pathways impinge on the environment at different places.  Environmental damage may occur at any point along the route, in three forms:

1.  damage to economically valuable goods (e.g., DDT in fish),
2.  human disease and mortality (e.g., Hg in fish and shellfish eaten by humans),
3.  changes in physical environment (e.g., climatic changes owing to increased atmospheric burden of particulates).

Unfortunately, not nearly enough is known about the damage done by pollutants.  Data are available on which to make assessments of rates of pollutant inputs into the environment, their circulation through the environment and ultimate fate. However, acute and chronic toxcities of pollutants are poorly known, as are their effects on the physical environment.

There are several general factors (Table 1) whose continued future growth will undoubtedly increase problems of environmental degradation, unless measures are taken to slow their growth or to alleviate their impact on the environment in other ways.

These factors are:

L. Population: Most estimates of world population for the year 2000 are about 7 billion, even assuming reduction in birth rates. One basic cause (even if birth rate is drastically decreased) is increased longevity. Continued decline in death rates and progress toward equality of births and deaths would give level population of about 15 billion late in the 21st century. It is anticipated that growth will be maximal in currently poor countries. Some authorities believe that population growth is the principal factor leading to environmental degradation. The increase of the U.S. population in urban areas from 40 to 70% during this century has lead to some serious localized pollutant discharges.

2. Gross National Product (GNP): As U.S. productivity per capita increases, gross national product can be expected to rise more steeply than population. The GNP/capita ratio of all industrial nations continues to rise along with energy demands, with accompanying increase in all types of residues -- solid wastes, atmospheric emissions, etc.

3. Energy: Energy demands are estimated to increase about 2% per capita per year for the next 20 years or so; increased population will mean total energy requirements growing at 3-4%/yr (Fig. 5). Usual projection is that world energy demands will increase 1/3 faster than U.S. demands. More and more coal, oil and gas will be used to generate electricity, rather than being used directly as fuel. Nuclear power will increase, but perhaps more slowly than estimated recently. Thus, even with low sulfur coal, pollution devices in cars, action by the oil producing nations of the Middle East, etc., we can expect emissions to increase. More energy will be required per unit of mineral resource produced. The cost of controlling emissions and wastes will increase. World supply of fossil fuel is shown in Table 2; the total of $540 \times 10^{17}$ kcal is equivalent to a few centuries of energy from fossil fuels. For the U.S.,

8

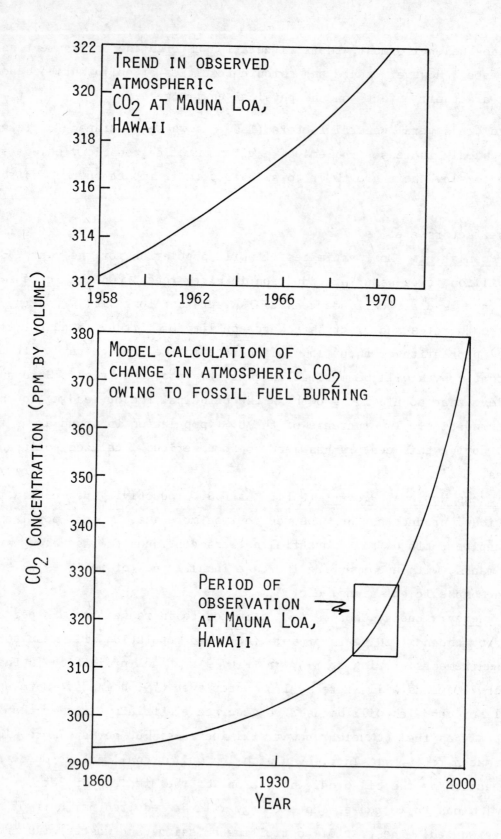

Figure 4. Carbon dioxide concentration in the atmosphere (from Machta, L., 1972, The role of the oceans and biosphere in the carbon dioxide cycle. in The Changing Chemistry of the Oceans, Dyrssen, D. and Jagner, D., eds., Wiley-Interscience, N. Y., 121-145.

Table 1

General Environmental Factors

| | | 1970 | 2000 (estimated) | Increase by Factor of |
|---|---|---|---|---|
| 1. | Population    World | 3.5 billion | 7.0 billion | 2 |
| | U.S. | 200 million | 300 million | 1.5 |
| 2. | U.S. GNP | $900 billion | $2000 billion | 2.2 |
| 3. | Electricity demand | 4000 trillion kcal | 23000 trillion kcal | 6.0 |
| | U.S. total energy demand | 15000 trillion kcal | 44000 trillion kcal | 3.0 |
| | Nuclear power | 30 trillion kcal | 10000 trillion kcal | hundreds |
| 4. | U.S. total emissions (CO, particulates, sulfur oxides, hydrocarbons, nitrogen oxides) | 200 million metric tons | 1200 million metric tons | 6.0 |
| 5. | World food production | 8 trillion kcal/day | 20 trillion kcal/day | 2.5 |
| 6. | U.S. water use | 1700 trillion $cm^3$/day | 3400 trillion $cm^3$/day | 2.0 |

(Note:  1 KWH = 860 kcal;  in U.S. one person consumes an average of 2500 kcal/day = 3 KWH)

the estimates usually are in the range of 100-150 years. Recently the U.S.
Geological Survey reduced the estimate of world fossil fuel by 1/3 to 1/2.
Because of its abundance, coal must bridge the gap between short-term and
long-term solutions to the energy problem. Uranium resources for breeder
reactors depend on price: $7.6 \times 10^{20}$ kcal (at \$10/lb $U_3O_8$), or as much as
$12600 \times 10^{20}$ kcal if we are willing to pay \$500/lb to recover low grade ore.
The energy content of these fuels far exceeds that of fossil fuels. Use
of these fuel sources can lead to environmental problems such as $SO_2$ release from burning coal, disposal of radioactive residues from nuclear plants.

4. Emissions: The estimate of emission increase is by a factor of 6
or so. This is probably high; strong current efforts to control emissions
may reduce the rate of increase. The greatest problem is release in urban
areas where emissions can become very concentrated. Cities cover about 2%
of the U.S. The questions we will consider are: What happens to emissions
after they leave the urban area? Where do emissions go and what effects do
they have?

5. World Food Production: World food production is 8 trillion kilo-
calories/day, or about 2300 kcal/person/day. If evenly distributed, this would
provide a minimal number of calories for each person in the world. Any pro-
jections, based on the hope of minimal adequate nutrition, are of the order
of 20 trillion kilocalories/day with a doubled population. Such production does
not seem impossible, but there are serious worries about the ability to produce enough fertilizer, the effects of the fertilizers required, about equal
distribution of food (as now), and about the restriction of numbers of species
of plants used for food (because of the possibility of blights). Food production is tied to energy requirements. In primitive cultures 1 kilocalorie
of energy produced 5-10 food kilocalories. Today in industrialized societies,
25-50 kilocalories of energy are expended to produce 5-10 food kilocalories.
So although mechanized agriculture with lavish use of fertilizers increases
the world food crop, we are paying a high price in energy consumption. Protein is essential for human diets, but we have become consumers of larger and
larger quantities of animal protein, which requires considerably more energy
to produce than does grain protein.

6. Water Use: Fresh water use (withdrawals) in the U.S. will approach
total stream flow by the year 2000, and is nearly half now. Water consumption,

Table 2

Energy Content of World Supply of Fossil Fuel

(units of $10^{17}$ kcal)

|  | Eventually recoverable[a] | % total |
|---|---|---|
| coal | 480 | 88.8 |
| crude oil | 28 | 5.2 |
| natural gas | 25 | 4.7 |
| tar-sand oil | 4.3 | 0.8 |
| shale oil | 2.8 | 0.5 |
| Total | 540.1[b] | |

a From M.K. Hubbert, 1970. Energy resources for power production. Proc.
IAEA Symp. Environmental Aspects of Nuclear Power Stations, N.Y., Aug.
1970. IAEA-SM-146/1.

b McKelvey and Duncan estimate a more optimistic value of $1,134 \times 10^{17}$ kcal.
(McKelvey, V.E. and Duncan, D.C., 1967. United States and world resources
of energy. Proc. 3rd Symposium of the Development of Petroleum Resources
of Asia and the Far East, UN-ECAFE, Mineral Resources Development Series,
26, v. 2)

---

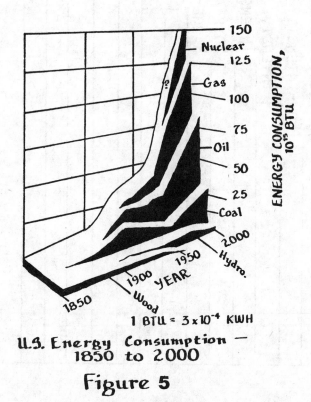

U.S. Energy Consumption —
1850 to 2000

Figure 5

Adapted from Batelle Research Outlook, 1972,
Our energy supply and its future, 4, 40 p.

12

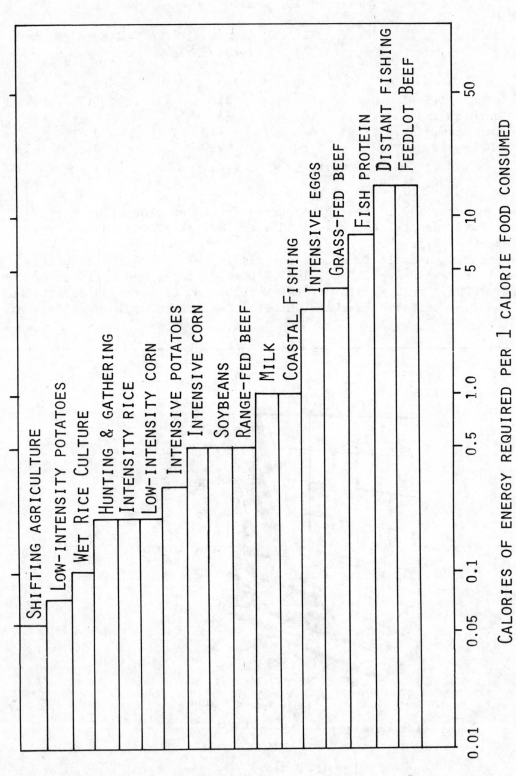

Figure 6. Energy requirements for various foods (adapted from Steinhart, J. S. and C. E. Steinhart, 1974. Energy use in the U.S. food system. Science, 184, 307-316.

which means removal from use, is still low.  Most consumption is by irrigation evaporation;  some is by spoilage due to added nutrients or industrial chemicals.  Water for the future is largely a local problem in dry areas.  Most U.S. streams are in good condition, with the exceptions closely tied to local sewage, irrigation, and industrial pollution.  By 2000, water consumed on a global basis may have increased by 65%, and consumption could amount to nearly one-third of the entire runoff.

---

Selected References:

Batelle Research Outlook, 1972.  Our Energy Supply and its Future, 4 (1), 40 pp.
Brubaker, S., 1972.  To Live on Earth, The Johns Hopkins Press, Chapter 2.
Häfele, W., 1972.  A systems approach to energy, Amer. Sci., 62, 438-447.
National Academy of Sciences, 1972.  Materials and Man's Needs, Wash., D.C.,
        217 pp.
Ridker, R.G., 1972.  Population and pollution in the United States, Science
        176, 1085-1090.
Science, 188 (4188) 1975. (Entire issue on food).

14

1. What are the three chief fossil fuels?  List in order of importance of world reserves.

2. What are the chief emissions to the atmosphere from the burning of fossil fuels?

3. Gasoline is very low in nitrogen.  Why then are automobiles a major source of addition of nitrogen oxides to the atmosphere?

4. If reasonable predictions can be made for world population and world energy production in the year 2000, why are predictions of man's additions to the earth's surface environment so unreliable?

5. Extensive development of nuclear power presumably would reduce markedly the problem of fossil fuel emissions.  At present there is little release of radioactive nuclear wastes to the environment.  Why then is there much concern about expansion of nuclear power plants?

6. Why is there much current emphasis on development of the "breeder reactor"?

7. Why not develop wind, tidal, and water power, as well as solar energy, especially because they are non-polluting?

8. How about energy from nuclear fusion processes?

9. Total energy consumption in 1975 for the U.S. is estimated at $2.33 \times 10^{13}$ KWH.  (a)  How many kilocalories does this represent?  (b)  How much energy do we use per capita compared to that in the food we eat?  (Population of U.S. 1975 approximately 210 million.)

10. If the U.S. uses 30% of total world energy production, what is the per capita energy consumption worldwide?

11. What is the world-wide ratio of total energy production to food energy, using 2000 kcal/day/person as the present world food consumption?

12. If energy production per capita is a measure of a desired standard of living, what would be current world energy production if everyone consumed as much as we do in the U.S.?

13. Even if population remained unchanged in the U.S., as well as life style, give arguments to show that energy consumption per capita would probably still increase.

14. If a strong program were developed for home use of solar and wind energy, what would the maximum effect be on centralized (and pollutional) energy production?

15. The following figure illustrates the cycle of iron and steel in the U.S. Wastes account for what percentage of total steel production?  Where do these wastes go in the environment?  Total U.S. emissions in 1970 were 200

million metric tons;  what proportion of these emissions was particulate
iron generated by U.S. iron and steel industrial operations?

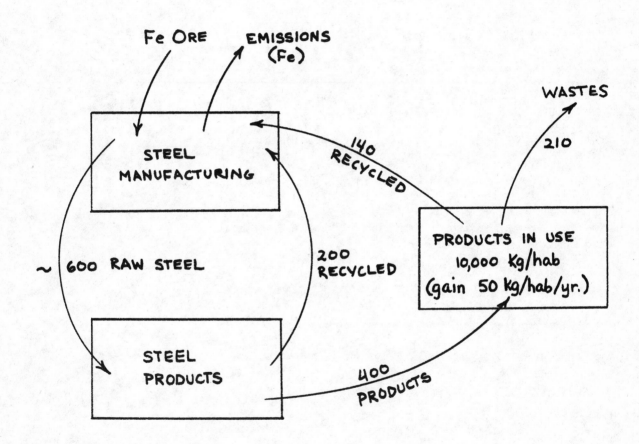

The cycle of iron and steel in United States society.
Fluxes are in units of kg/habitant/year. (written
communication, R. Wollast, 1974).

**NOTES**

Chapter 2


RESIDENCE TIME


## Introduction

The term residence time is widely used in many kinds of environmental studies. In modeling dynamic systems of many types in which there is continuous transfer of materials from one place to another, the most common approach is to divide the system into a series of storage places, or reservoirs, and to study the flow of substances into and out of the arbitrarily defined reservoirs. The flow rate in or out is commonly called the flux of the substance. Residence time is then defined as

$$\frac{\text{amount of material in the reservoir at a given time}}{\text{instantaneous rate of addition (or subtraction) of material}}.$$

Residence times, then, are defined at a chosen moment in time. Several examples may help to show some of the many ways in which residence times are calculated and used.

## Constant reservoir, constant flux in, constant flux out, single source and sink

If a bathtub is filled to a given level, containing 50 gallons of water, the plug is removed, and the tap is turned on so that the water level does not change, a situation is achieved in which there is a constant flux of water into the tub, which equals the flux out the drain. If the rate of water flow from the tap is measured at 5 gallons per minute, the residence time of water in the tub at any given moment is

50 gallons/5 gallons/minute = 10 minutes.

The preceding situation, with constant reservoir size, constant and equal influx and efflux, is called a steady state. As will be seen, an approach to this condition is achieved in many natural situations.

If the rate from the tap were suddenly increased to 6 gallons/minute, and held constant at that rate, while the amount going down the drain remained constant at 5 gallons/minute, the residence time would continuously increase, and it would be different with respect to influx and efflux. Twenty minutes after the change in rate of influx, the tub would contain 70 gallons of water; formally, [50 + 6 x 20 - 5 x 20 = 70]. The residence time for the water in the tub, with respect to influx, at the end of the 20 minutes would be

70 gallons/6 gallons/minute = 11.7 minutes,

whereas its residence time with respect to efflux would be

70 gallons/5 gallons/minute = 14 minutes.

Note that the same change in rate of influx, if the original reservoir had been much larger, would produce much smaller percentage changes in residence times.

Residence time is sometimes identified incorrectly with renewal time. For the steady-state bathtub, with 50 gallons of stored water and 5 gallons per minute influx and efflux, the residence time of water of 10 minutes in the tub is obviously the time required to add or subtract an amount of water equal to that in the tub. Does this mean that original, dirty water in the tub would be completely replaced by clean water from the tap in 10 minutes? The answer could possibly be yes, but in most real situations is no. A condition that would completely renew the water would be one in which the new, clean water did not mix at all with the old, dirty water, but simply displaced it out the drain. One physical situation that would approach this ideal would be to have the tub full of warm, dirty water, and to bring in clean, cold, dense water at the bottom, thus pushing the old water out of an overflow at the top of the tub. The reverse of this situation could also occur; if the faucet were placed directly above the drain, a situation could be imagined in which new clean water disappeared directly down the drain, leaving the amount of dirty water in the tub unchanged.

In many natural reservoirs an intermediate situation occurs, one represented in the bathtub example by continuously and thoroughly mixing the new, clean water with the old, dirty water. At the end of 10 minutes the water in the tub would be a 50-50 mixture of new and old water, so the tub water could be said to be half purified. At the end of a second ten minutes, 3/4 of the water would be clean, new water, and so on. For this complete mixing situation, the percentage of original water would diminish logarithmically, with half of the remaining dirty water removed every ten minutes. Just as for the decay of

radioactive substances, a <u>half-life</u> of ten minutes could be assigned to the
original water. At the end of an hour, or 6 half-lives, of this "complete
mixing steady state", the fraction of original water would be

$$1/2 \times 1/2 \times 1/2 \times 1/2 \times 1/2 \times 1/2 = 1/64.$$

It is common practice to regard 1/64 as an insignificant remaining fraction,
and to say that in the complete mixing situation, 6 half-lives or 6 residence
times are sufficient to <u>renew</u> the reservoir.

Thus the relations between residence time and mixing are critical in real
problems. If mixing is slow compared to influx and efflux, new and old water
have highly variable ratios in different parts of the reservoir. If mixing is
fast compared to residence time, new and old water are uniformly intermingled.
The same argument applies to the mixing of dissolved substances in the new or
old water; if the new water entering a lake contains high concentrations of
mercury, for example, the distribution of mercury in the lake will depend on
fluxes, reservoir size, and mixing. If a dissolved substance is found to be
uniformly distributed in a reservoir, it can be concluded that the residence
time is much longer than that of mixing.

## Multiple Fluxes

Many lakes approximate the steady-state bathtub example. Lake Erie main-
tains a nearly constant level, and has done so for centuries. Like the bathtub, it is fed by the 'faucet' of the Detroit River, and drained by the Niagara
River. In addition to these fluxes, water comes into the lake from other smaller
river sources, from precipitation, and from underground waters percolating
through the rocks of the lake basin. It also loses water by evaporation. For
Lake Erie the <u>total</u> <u>residence</u> <u>time</u> of the water, over a period of constancy,
or near-constancy of lake level, is equal to the total water in the lake/influx
rate from all sources, and also equal to the total water in the lake/efflux
rate from all removal mechanisms.

On the other hand, it is often useful to calculate the residence time with
respect to only one of the several fluxes. For example, what is the residence
time for the water of Lake Erie with respect to the input from the Detroit River?
Values for residence times related to individual fluxes are a convenient way
of comparing the importance of the various inputs or outputs. A short residence time for one input, versus a long one for another, immediately focuses
attention on the former.

## Reservoirs within Reservoirs

In studies of the cycling of materials through the earth's surface environment, there is a tendency to make a gross classification of reservoirs into the atmosphere reservoir, the ocean reservoir, the rock reservoir, and the biologic reservoir. These reservoirs are then subdivided as convenient into "the nitrogen reservoir in the atmosphere", the "calcium reservoir in the ocean", "the uranium reservoir in rocks", "the nitrogen reservoir in biologic materials", and so on. In the jargon of the environmentalist, these subdivisions may be termed compartments. The reservoir-flux concept can be applied arbitrarily at any desired scale, depending on the problem. A household, with its multiple input fluxes, can be designated as a reservoir, for purposes of studying energy consumption and release, perhaps to assess the best method of lowering consumption of energy or eliminating waste.

## Residence Time and Pollution

Residence time calculations are widely used because they often give definitive answers to important problems, or at least suggest the most important aspects of the problem to be investigated. A few years ago suggestions were made that the burning of fossil fuels would consume so much of the oxygen in the atmosphere that life on earth would be endangered. In a brilliant article, Broecker (Man's oxygen reserves, W.S. Broecker, in Science 168, 1537-1538, 1970) showed that the residence time of oxygen in the atmosphere, at any conceivable removal rate by fossil fuel burning, and with no influx at all, would be millions and millions of years. The rate of removal, compared to the reservoir, was found to be infinitesimally small. To clinch the argument, he demonstrated that if all reserves of fossil fuels were burned instantaneously, the percentage change of oxygen in the atmospheric reservoir would be trivial.

On the other side of the coin, demonstration that a pollutant has a very short residence time in the atmosphere or other reservoir may be almost enough evidence to dismiss it as a long-term cumulative peril. Carbon monoxide has a residence time in the atmosphere, calculated from the total reservoir and total input rate (including man's activities), of about 3 weeks. This number suggests that carbon monoxide has a rate of removal of about 3 weeks as well; otherwise the amount in the reservoir would increase at an easily observable rate. If no carbon monoxide were removed or converted to other compounds, the atmospheric content would double in about 3 weeks; in one year global

levels would rise by a factor of 52/3, or about 17-fold.  Because the global
level has not changed measureably for a number of years, it can be concluded
that removal of carbon monoxide must be keeping pace with production.

## The Earth System

It will be shown that over long periods of time, at least for hundreds of
millions of years, the cycling of materials through the earth's surface environment has approximated a steady-state system.  This statement applies especially
to the ocean and atmosphere.  The argument is based largely on the fact that if
the cycling were not a close approximation to a steady state, with input to the
atmosphere and ocean closely matched by output, the resultant changes in composition of these two important reservoirs would have made the known fossil record
impossible.  The present-day complex assemblages of organisms are the evolved
descendants of hundreds of millions of years of continued existence of similarly
complex assemblages.  The patterns of evolution of organisms are not by any
means those that would be expected from constant conditions of atmosphere and
ocean.  There are clearly times of increase and decrease of environmental stress.
But neither ocean nor atmosphere has changed enough to prevent a complex biota
from existing continuously.  The exact permissible ranges of conditions of composition and temperature of the oceans, or of oxygen, carbon dioxide, and other
atmospheric gases cannot be pinpointed, but most scientists would agree that a
man transported back 300 million years in a time machine would find water to
drink and food to eat and air to breathe that would permit him to survive.

**NOTES**

Chapter 3


THE ATMOSPHERE


## Composition, Structure and Circulation

The area of the earth is $5.1 \times 10^{18} cm^2$, or $197 \times 10^6$ square miles. The area of the oceans is about 70% of the total, or $3.6 \times 10^{18} cm^2$; that of the land $1.5 \times 10^{18} cm^2$. The average height of the land is 0.8 km (1/2 mile), or $0.8 \times 10^5$ cm; the average depth of the oceans $3.8 \times 10^5$ cm (2.3 miles). Greatest land heights are about 30,000 feet; greatest ocean depths about 35,000 feet.

The mass of the atmosphere is $52 \times 10^{20}$ g; it can be obtained from the weight of air above 1 $cm^2$ of earth's surface and total earth area:

$$1,031 g/cm^2 \times 5.1 \times 10^{18} cm^2 = 52 \times 10^{20} g.$$

Air is made up chiefly of oxygen ($O_2$) and nitrogen ($N_2$), with molecular weights of 32 and 28, respectively. The average molecular weight of air is 29g, so the total number of moles of gas is $52 \times 10^{20} g/29 g/mole = 1.8 \times 10^{20}$ moles. The moles of $O_2$ are $0.2094 \times 1.8 \times 10^{20} = 0.38 \times 10^{20}$; the moles of $N_2$ are $0.7807 \times 1.8 \times 10^{20} = 1.41 \times 10^{20}$. Carbon dioxide is a minor but important constituent; total moles are $0.000314 \times 1.8 \times 10^{20} = 0.00056 \times 10^{20}$. Atmospheric composition is given in Table 3.

Pressure-height relations are shown in Figure 7. Pressure halves every 5 km; thus 1/2 the atmospheric mass is under 5 km. When you fly in a jet at 10 km, 3/4 of the mass of the atmosphere is below you.

Temperature-height relations are shown in Figure 8. The average temperature for the whole earth's surface is 15°C. Temperature diminishes nearly linearly up to about 12 km, then increases again slowly, reaching about 0°C at 50 km, it then declines to about -80°C at 80 km, rises again at still higher elevations. The temperature maximum at 50 km is caused by absorption of ultraviolet light by oxygen to make ozone. The atmosphere is heated from below. It is nearly transparent to sunlight. But when sunlight reaches the earth's surface the visible wavelengths are absorbed, the earth's surface is heated, and it reradiates longer wavelengths (infrared). These infrared rays are absorbed by water vapor and especially by $CO_2$, warming the atmsophere adjacent to the earth's surface.

Warm air rises, then cools by expansion. Cool surface air moving in to take the place of rising warm air causes winds. Thus the lower part of the atmosphere (troposphere) is unstable and tends to circulate and mix. Above the troposphere, where temperature rises again, the atmosphere is more stable vertically. Difference in mixing of troposphere and stratosphere is well illustrated by a smoke plume from a chinmey which dissipates quickly, versus persistent jet trails.

Substances that remain for a long time in the atmosphere tend to be well-mixed throughout the atmosphere. Carbon dioxide is a good example (Fig. 9). Its proportion in dry air, except for seasonal variation of about 10 ppm, is nearly constant. Ozone, on the other hand, is generated in the stratosphere, then mixes downward into the troposphere and is converted back to oxygen. Carbon monoxide shows a reverse profile; it is generated in the lower atmosphere and oxidized to $CO_2$ in the upper atmosphere. Water vapor, with a similar profile, is frozen out of cold high air.

The mass of air in a given volume diminishes logarithmically with height; the mass of water vapor per given volume diminishes much more rapidly, so the stratosphere has a very low water content (hence part of the concern about supersonic planes and what they will put into the high atmosphere). The dust maximum in the stratosphere (Fig. 10) comes chiefly from volcanic ejections (and atomic bombs). Of great importance are the mixing times of the various parts of the atmosphere; a dust particle may reside several years in the stratosphere before the layered structure permits it to get back into the troposphere, where mixing is rapid.

Atmospheric circulation is sketched roughly in Figures 11 and 12. Maximum heating takes place where the sun is overhead, on the average at the equator. Warm air rises from the ocean surface, cools, moves N and S at high elevation. Cooler air moves in from N and S to take its place, creating the trade winds. The Doldrums are an area of high rainfall and cloudiness. As the trades blow toward the equator they are veered as the earth rotates from W to E beneath them (Coriolis effect), and they flow from a place of lower rotational speed to a higher one. As the trades flow toward the equator they are warmed, their relative humidity drops, and they pick up moisture from the ocean.

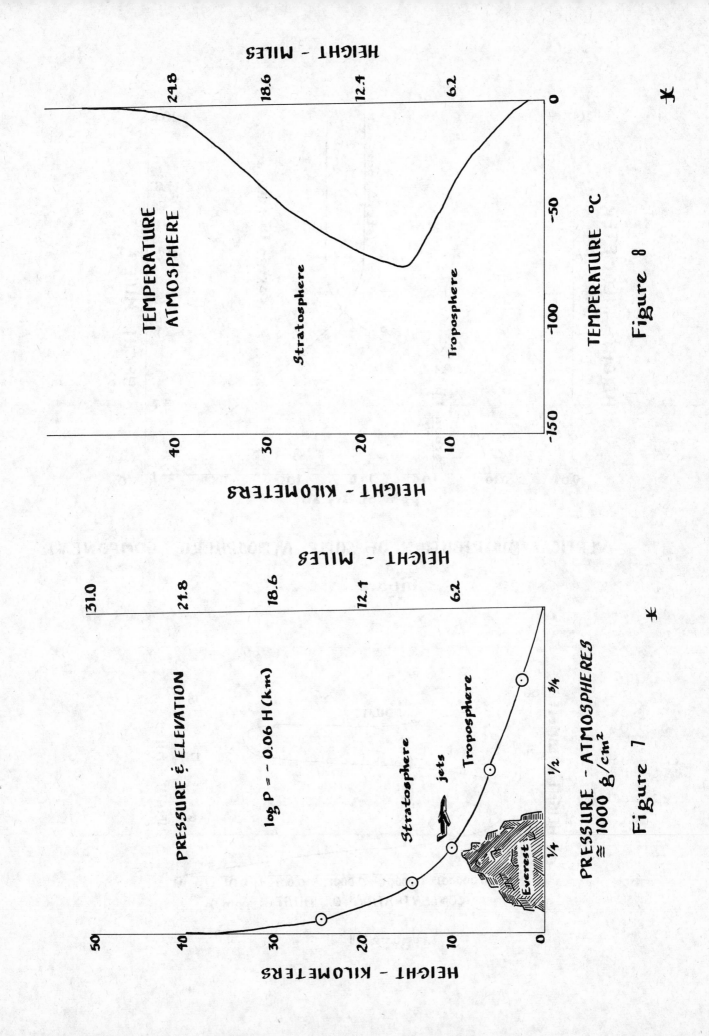

HEIGHT – MILES

24.8    18.6    12.4    6.2

TEMPERATURE
ATMOSPHERE

Stratosphere

Troposphere

TEMPERATURE °C

-50    -100    -150

Figure 8

40    30    20    10

HEIGHT – KILOMETERS

HEIGHT – MILES

31.0    24.8    18.6    12.4    6.2

PRESSURE & ELEVATION

log P = – 0.06 H (km)

Stratosphere
jets
Troposphere

Everest

PRESSURE – ATMOSPHERES
≅ 1000 g/cm²

¼    ½    ¾

Figure 7

50    40    30    20    10

HEIGHT – KILOMETERS

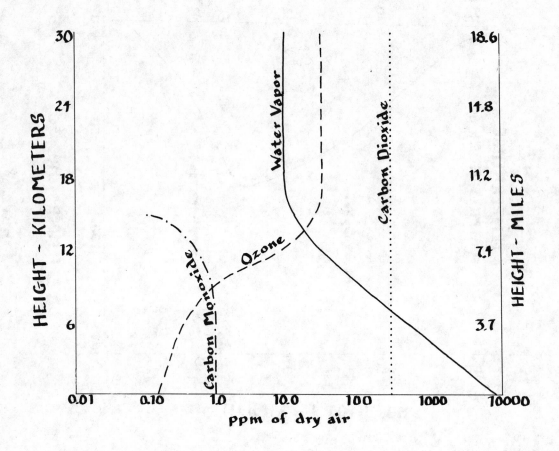

VERTICAL DISTRIBUTION OF SOME ATMOSPHERIC COMPONENTS

Figure 9

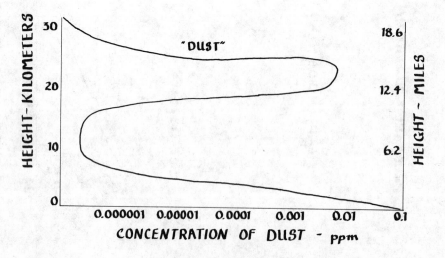

Figure 10

JC

Part of the air descending from the Doldrums flows N or S starting at about 20°. It is veered to the E. Encounter between this air and polar air creates whirling air masses in the troposphere hundreds of miles across that migrate to the E at 20-30 km/hr -- the Westerlies. Most of the world's industry is in the belt of the Westerlies. Atmospheric pollutants produced in this belt move E.

Figure 13 shows wind velocities as a function of latitude, altitude, and season. Positive numbers on the figure are for W winds, negative for E. Note that in general velocity increases upward, with velocities reaching 130 km/hr (70 mph). The tendency is for a velocity minimum at the top of the troposphere, a general velocity increase upward in the stratosphere, and for seasonal change in velocities in N and S hemispheres.

The residence time of water vapor (0.2 wt % of atmosphere) is only about 10 days, i.e., total precipitation for 10 days = mass of $H_2O$ in atmosphere. The fate of many pollutants is related to "rain out", thus the small residence time of $H_2O$ is a major control of atmospheric pollution.

## Pollution and the atmosphere

None of the gases in the atmosphere can be regarded as a permanent resident. Oxygen is continuously generated by photosynthesis and is removed by respiration and decay. It is also continuously being dissolved in and emitted from the oceans, and is used in the oxidation of inorganic substances on land (ferrous compounds and sulfide compounds). But it takes about 7,600 years for atmospheric $O_2$ to be completely cycled through plants. $N_2$ is similar to $O_2$ in its general behavior, but its annual use and return is smaller than that of $O_2$, so its turnover time is even longer. The circulation of $CO_2$ through plants takes only about 10 years. For these three constituents, the residence time is long compared to complete mixing time (a few years), so their relative percentages are almost constant throughout the atmosphere.

The gases below hydrogen in Table 3, many of which are considered pollutants, are also added and subtracted by natural processes, and for many of them the amounts added and subtracted per year are comparable to the total in the atmosphere, so concentration gradients develop, as shown in Figure 9. The natural sources of the gases from the bottom of Table 3 up to hydrogen (except $O_3$) are largely decay processes. These gases are used up by oxidative processes. Knowledge of the magnitudes of their sources is very uneven.

Table 3

Composition of the Atmosphere

| Constituent | Mol % | Molecular Weight | No. Moles |
|---|---|---|---|
| Nitrogen ($N_2$) | 78.084 | 28.0134 | $1.41 \times 10^{20}$ |
| Oxygen ($O_2$) | 20.9476 | 31.9998 | $0.38 \times 10^{20}$ |
| Argon (Ar) | 00.934 | 39.948 | $0.0168 \times 10^{20}$ |
| Carbon dioxide ($CO_2$) | 00.0314 | 44.0095 | $0.00056 \times 10^{20}$ |
| Neon (Ne) | 00.001818 | 20.183 | $3.27 \times 10^{15}$ |
| Helium (He) | 00.000524 | 4.0026 | $0.94 \times 10^{15}$ |
| Krypton (Kr) | 00.000114 | 83.80 | $0.21 \times 10^{15}$ |
| Xenon (Xe) | 00.0000087 | 131.30 | $1.57 \times 10^{13}$ |
| Hydrogen ($H_2$) | 00.00005 | 2.01594 | $9.0 \times 10^{13}$ |
| Methane ($CH_4$) | 00.0002 | 16.04303 | $36.0 \times 10^{13}$ |
| Nitrous oxide ($N_2O$) | 00.00005 | 44.0128 | $9.0 \times 10^{13}$ |
| Ozone ($O_3$) | 0 to 00.00007 | 47.9982 | variable |
| Sulfur dioxide ($SO_2$) | 0 to 00.0001 | 64.0628 | variable |
| Nitrogen dioxide ($NO_2$) | 0 to 00.000002 | 46.0055 | variable |
| Ammonia ($NH_3$) | trace | 17.03061 | variable |
| Carbon monoxide (CO) | trace | 28.01055 | variable |
| | | Total moles | $1.81 \times 10^{20}$ |

Adapted from K.K. Turekian, The Oceans, Streams, and Atmosphere. in Handbook of Geochemistry, vol. 1, K.H. Wedepohl, ed., Springer-Verlag, New York, 1969, p. 318.

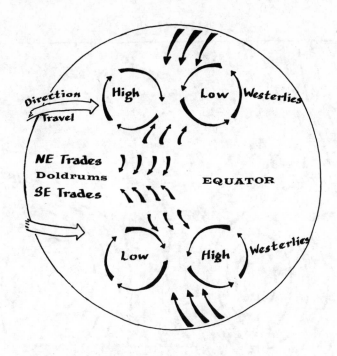

PLAN VIEW OF ATMOSPHERIC
CIRCULATION

Figure 11

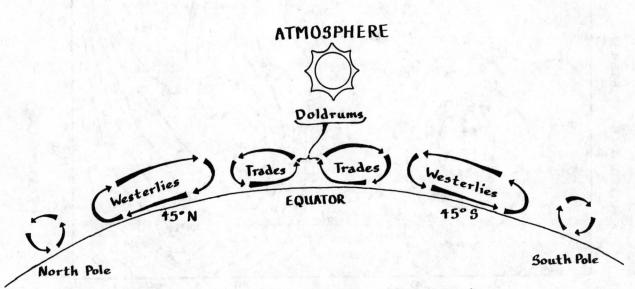

Schematic Vertical Circulation of Atmosphere

Figure 12

# WIND VELOCITIES AS A FUNCTION OF
## LATITUDE, ALTITUDE & NORTHERN HEMISPHERE SEASON

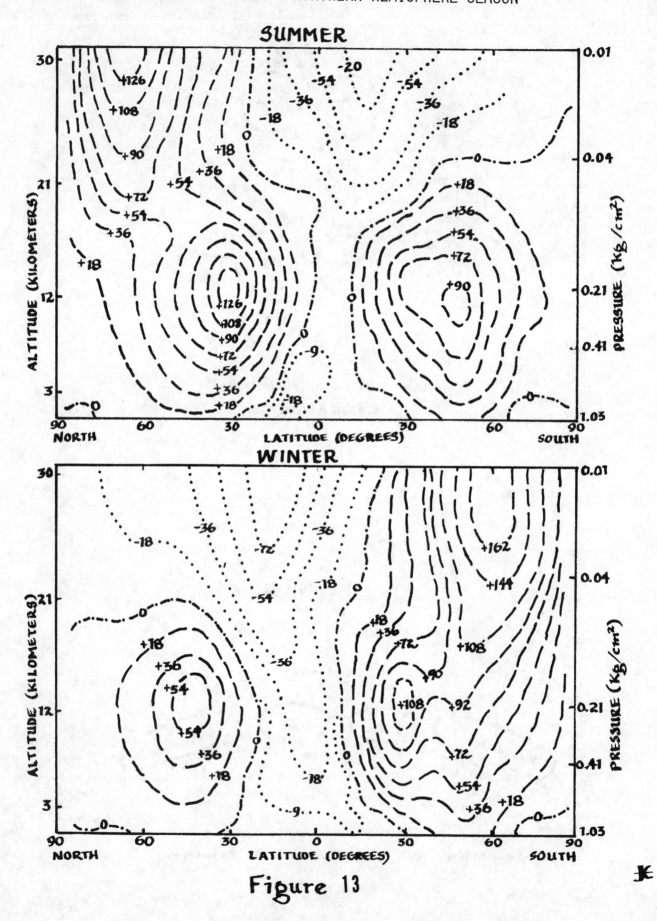

### Figure 13

ADAPTED FROM NEWELL, R.E., 1971, THE GLOBAL CIRCULATION
OF ATMOSPHERIC POLLUTANTS. SCIENTIFIC AMERICAN, 224, 32-42.

Decay and respiration can be regarded as low temperature combustion of organic matter. The burning of fossil fuels, at much higher temperatures, produces similar gases. Just as fossil fuel burning produces gases both reduced and oxidized, depending on combustion conditions, so natural low temperature combustion produces a comparable range. Also, because the materials combusted by both processes have similar chemical compositions, the proportions of the elements released to the atmosphere are broadly similar.

Study Questions

1.  If atmospheric pressure at sea level is 1.031 kg/cm$^2$, what is the total mass of the atmosphere?

2.  If the average molecular weight of the gases of the atmosphere is 29, what is the total number of moles?

3.  According to Table 3, nitrogen makes up 78.084 mol % of the atmosphere. How many moles of $N_2$? How many grams?

4.  If warm moist air contains about 4 mol % $H_2O$, what is its density relative to dry air at the same temperature?

5.  If air is warmed, its volume increases proportionally to the increase in absolute temperature (°K = °C + 273). If air is warmed from 25°C (77°F) to 35°C (95°F), what is its % change in density?

6.  Summarizing #4 and #5, a 10°C change in temperature, plus an increase of water content of 4 mol %, results in a decrease of density of about 5% – 2% from addition of water vapor, 3% from increase in temperature. What is the application of this information to the air circulation in the Trade Winds and Doldrums?

7.  If a rocket at the North Pole shoots a space missile at a surfaced submarine at the equator at velocity of 6,000 miles per hour, and shoots directly down the latitude line connecting the missile site and the submarine, will it hit or miss the submarine? If it will miss, by approximately how much?

8.  Why is it said that the troposphere is unstable, and the stratosphere stable?

9.  Why do troposphere temperatures decrease upwards?

10. Why does the stratosphere have a positive temperature gradient?

11. What are temperature inversions, and why are they so important in local pollutional problems?

12. What should be the most efficient routes for jets between San Francisco and Hawaii?

13. Water vapor in atmosphere is about 0.2 weight %. What is the total weight of water vapor in atmosphere?

14. Total evaporation is 4.46 x 10$^{20}$ g/yr. What is the residence time of water vapor in the atmosphere?

15. What is the approximate time necessary for air to circle the globe at 45°N latitude in summer?

16. If a pollutant is susceptible to being rained out of the atmosphere, would you expect, from questions 13, 14 and 15, that it would mix evenly throughout the troposphere?

Chapter 4


THE OCEANS


## Composition, Structure and Circulation

The mass of the oceans is about 14000 x $10^{20}$ g, or 270 times that of the
atmosphere.  The average pressure exerted on the ocean floor is about 400 times
that exerted by the atmosphere on the ocean surface.  Ocean water contains 3.5
wt % dissolved salts, and has a density of about 1.026g/cm$^3$, as compared to 1.0
for pure water.  The total salt mass is about 500 x $10^{20}$ g (10 x the mass of the
atmosphere).

Distribution of land and sea is 70% sea and 30% land;  the northern hemisphere is about 50% land, the southern about 25%.  Land distribution by latitude
is shown in Figure 14.  Note the high percent of land in the belt of the Westerly
winds, and the absence of land in the latitudinal belt at 60°S, which permits
circumglobal oceanic circulation in that region.

The ocean is a squashed mirror image of the atmosphere.  The atmosphere is
heated at the bottom, causing instability and mixing in the troposphere;  the
ocean is heated at the top, resulting in general stability.  Rapid mixing occurs
only in shallow water as the result of wave and current action activated by winds.
Figure 15 shows temperature relations in atmosphere and ocean, and suggests the
mixing relations as well.

The composition of the dissolved major species in seawater is given in Table
4, and a complete list of estimates of all 92 elements in Table 5.  Seawater is
dominated by $Na^+$, $Mg^{++}$, $Cl^-$ and $SO_4^=$.  Table 5 shows that 70 elements have been
detected in seawater, although only 7 chemical species make up more than 99% of
the total dissolved solids.

Figure 16 gives vertical profiles of some important minor constituents of
the ocean -- oxygen, phosphorus and nitrogen.  In contrast to the major species,
which remain almost constant vertically, these substances vary in concentration
with depth.  In the well-mixed, lighted surface waters, microscopic plants and
animals abound.  The plants (phytoplankton) photosynthesize $CO_2$, $H_2O$, N, S and
P into their cells;  in turn they are fed upon by animals who also use these

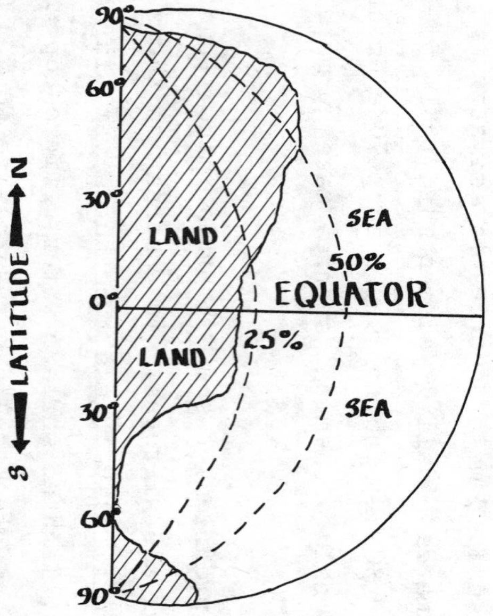

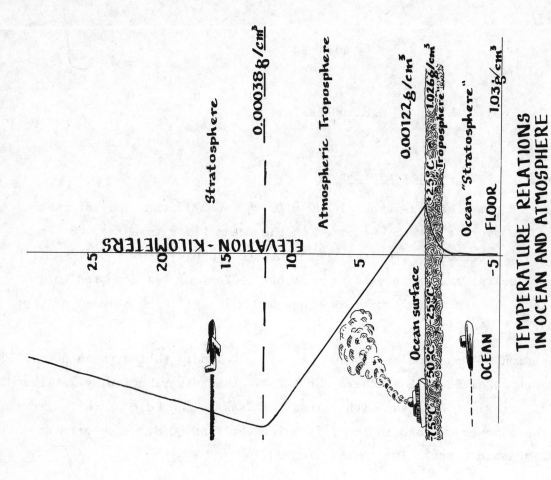

Stratosphere

0.00038 g/cm³

Atmospheric Troposphere

ELEVATION - KILOMETERS

25    20    15    10    5                    -5

0.00122 g/cm³    1.026 g/cm³

+2.5°C    Troposphere

-25°C

Ocean surface

-50°C

Ocean "Stratosphere"

-75°C    FLOOR    1.03 g/cm³

OCEAN

TEMPERATURE RELATIONS
IN OCEAN AND ATMOSPHERE

Figure 15

SEA

50%

EQUATOR

SEA

25%

LAND    LAND

90°  60°  30°  0°  30°  60°  90°

N ◄—— LATITUDE ——► S

DISTRIBUTION OF LAND & SEA

Figure 14

ADOPTED FROM DEFANT, A., 1961, PHYSICAL OCEANOGRAPHY,
VOL. 1, P. 4, PERGAMON PRESS, NEW YORK.

Table 4

Major Dissolved Species in Seawater

| Constituent | ppm (mg/liter or µg/g) |
|---|---|
| $Cl^-$ | 19000 |
| $Na^+$ | 10500 |
| $Mg^{++}$ | 1300 |
| $SO_4^=$ | 2650 |
| $K^+$ | 380 |
| $Ca^{++}$ | 400 |
| $HCO_3^-$ | 140 |
| $SiO_2$ | 6 |
| $Br^-$ | 65 |
| $CO_3^=$ | 18 |
| $Sr^{++}$ | 8 |
| Al | 0.001 |
| B | 4.5 |
| $F^-$ | 1.3 |
| Dissolved organic (as C) | 0.5 |

-------------------------------------------

Table 5

Composition of Streams and the Ocean

| Atomic Number | Atomic Weight | Element | Seawater (µg/1) or (ppb) | Streams (µg/1) or (ppb) |
|---|---|---|---|---|
| 1 | 1.00797 | hydrogen | $1.0 \times 10^8$ | $1.10 \times 10^8$ |
| 2 | 4.0026 | helium | 0.0072 | * |
| 3 | 6.939 | lithium | 170 | 3 |
| 4 | 9.0122 | beryllium | 0.0006 | * |
| 5 | 10.811 | boron | 4450 | 10 |
| 6 | 12.01115 | carbon (inorganic) | 28000 | 11500 |
| | | (dissolved organic) | 500 | * |
| 7 | 14.0067 | nitrogen (dissolved $N_2$) | 15000 | * |
| | | (as $NO_3^{-1}$, $NO_2^{-1}$, $NH_4^{+1}$, and dissolved organic) | 670 | 226 |
| 8 | 15.9994 | oxygen (dissolved $O_2$) | 6000 | * |
| | | (as $H_2O$) | $8.83 \times 10^8$ | $8.83 \times 10^8$ |
| 9 | 18.9984 | fluorine | 1300 | 100 |

## Table 5 (continued)

| Atomic Number | Atomic Weight | Element | Seawater (μg/1) or (ppb) | Streams (μg/1) or (ppb) |
|---|---|---|---|---|
| 10 | 20.183 | neon | 0.120 | * |
| 11 | 22.9898 | sodium | $1.08 \times 10^7$ | 6300 |
| 12 | 24.312 | magnesium | $1.29 \times 10^6$ | 4100 |
| 13 | 26.9815 | aluminum | 1 | 400 |
| 14 | 28.086 | silicon | 2900 | 6100 |
| 15 | 30.9738 | phosphorus | 88 | 20 |
| 16 | 32.064 | sulfur | $9.04 \times 10^5$ | 5600 |
| 17 | 35.453 | chlorine | $1.94 \times 10^7$ | 7800 |
| 18 | 39.948 | argon | 450 | * |
| 19 | 39.102 | potassium | $3.92 \times 10^5$ | 2300 |
| 20 | 40.08 | calcium | $4.11 \times 10^5$ | 15000 |
| 21 | 44.956 | scandium | 0.0004 | 0.004 |
| 22 | 47.90 | titanium | 1 | 3 |
| 23 | 50.942 | vanadium | 1.9 | 0.9 |
| 24 | 51.996 | chromium | 0.2 | 1 |
| 25 | 54.9380 | manganese | 1.9 | 7 |
| 26 | 55.847 | iron | 3.4 | 670 |
| 27 | 58.9332 | cobalt | 0.05 | 0.1 |
| 28 | 58.71 | nickel | 6.6 | 0.3 |
| 29 | 63.54 | copper | 2 | 7 |
| 30 | 65.37 | zinc | 2 | 20 |
| 31 | 69.72 | gallium | 0.03 | 0.09 |
| 32 | 72.59 | germanium | 0.06 | * |
| 33 | 74.9216 | arsenic | 2.6 | 2 |
| 34 | 78.96 | selenium | 0.090 | 0.2 |
| 35 | 79.909 | bromine | 67300 | 20 |
| 36 | 83.80 | krypton | 0.21 | * |
| 37 | 85.47 | rubidium | 120 | 1 |
| 38 | 87.62 | strontium | 8100 | 70 |
| 39 | 88.905 | yttrium | 0.013 | 0.07 |
| 40 | 91.22 | zirconium | 0.026 | * |
| 41 | 92.906 | niobium | 0.015 | * |
| 42 | 95.94 | molybdenum | 10 | 0.6 |
| 43 | | technetium | (not naturally occurring) | |
| 44 | 101.07 | ruthenium | 0.0007 | * |
| 45 | 102.905 | rhodium | * | * |
| 46 | 106.4 | palladium | * | * |
| 47 | 107.87 | silver | 0.28 | 0.3 |
| 48 | 112.4 | cadmium | 0.11 | * |
| 49 | 114.82 | indium | * | * |
| 50 | 118.69 | tin | 0.81 | * |
| 51 | 121.75 | antimony | 0.33 | 2 |
| 52 | 127.60 | tellurium | * | * |
| 53 | 126.90 | iodine | 64 | 7 |
| 54 | 131.30 | xenon | 0.47 | * |
| 55 | 132.91 | cesium | 0.30 | 0.02 |
| 56 | 137.34 | barium | 21 | 20 |
| 57 | 138.91 | lanthanum | 0.0034 | 0.2 |
| 58 | 140.12 | cerium | 0.0012 | 0.06 |
| 59 | 140.91 | praseodymium | 0.00064 | 0.03 |
| 60 | 144.24 | neodymium | 0.0028 | 0.2 |

Table 5 (continued)

| Atomic Number | Atomic Weight | Element | Seawater (µg/1) or (ppb) | Streams (µg/1) or (ppb) |
|---|---|---|---|---|
| 61 | | promethium | (not naturally occurring) | |
| 62 | 150.35 | samarium | 0.00045 | 0.03 |
| 63 | 151.96 | europium | 0.000130 | 0.007 |
| 64 | 157.25 | gadolinium | 0.00070 | 0.04 |
| 65 | 158.92 | terbium | 0.00014 | 0.008 |
| 66 | 162.50 | dysprosium | 0.00091 | 0.05 |
| 67 | 164.93 | holmium | 0.00022 | 0.01 |
| 68 | 167.26 | erbium | 0.00087 | 0.05 |
| 69 | 168.93 | thulium | 0.00017 | 0.009 |
| 70 | 173.04 | ytterbium | 0.00082 | 0.05 |
| 71 | 174.97 | lutetium | 0.00015 | 0.008 |
| 72 | 178.49 | hafnium | <0.008 | * |
| 73 | 180.95 | tantalum | <0.0025 | * |
| 74 | 183.85 | tungsten | <0.001 | 0.03 |
| 75 | 186.2 | rhenium | 0.0084 | * |
| 76 | 190.2 | osmium | * | * |
| 77 | 192.2 | irridium | * | * |
| 78 | 195.09 | platinum | * | * |
| 79 | 196.97 | gold | 0.011 | 0.002 |
| 80 | 200.59 | mercury | 0.15 | 0.07 |
| 81 | 204.37 | thallium | <0.01 | * |
| 82 | 207.19 | lead | 0.03 | 3 |
| 83 | 208.98 | bismuth | 0.02 | * |
| 84-89 and 91 | | (thorium and uranium decay series elements: polonium, astatine, radon, fracium, radium, actinium and protactinium) | | |
| 90 | 232.04 | thorium | <0.0005 | 0.1 |
| 92 | 238.03 | uranium | 3.3 | 0.3 |

* No data or reasonable estimates available

Adapted from K.K. Turekian, The Oceans, Streams, and Atmosphere, in Handbook of Geochemistry, vol. 1, K.H. Wedepohl, ed., Springer Verlag, New York, 1969, pp. 297-323.

nutrients. Phytoplankton deplete the surrounding water of nutrients, holding them at very low levels. With death and subsequent sinking of the organisms, oxidation by dissolved oxygen takes place, causing the oxygen minimum and releasing N and P to the surrounding water. Silica is another important nutrient for major types of phytoplankton, and shows similar vertical profiles (Fig. 17). Because phytoplankton are the base of the food chain (from plants to small animals to larger ones), the area of the oceans most important to man is the uppermost zone where phytoplankton are numerous. Oceanic productivity depends on the rate of return to the surface water of 'lost' nutrients. Also, because of slow mixing with deeper water, the shallow photic zone is most vulnerable to toxic additions, which may be added faster than mixing with bulk ocean water can take place, despite the huge dilutional capacity of the ocean as a whole.

Addition of nutrients to the ocean can increase the biomass, and often changes drastically the types of organisms present in a given region. Such 'fertilization' can be bad or good, according to man's wishes and needs. The richest fishing grounds in the world are off the coast of Peru, where favorable winds cause water circulation that brings deep nutrient-rich water to the surface. Experiments are in progress to accomplish the same thing elsewhere (e.g., U.S. Virgin Islands) by pumping up deep water, or by direct addition of nutrients to surface water. As we shall see, excess nutrient addition to lakes and estuaries is usually accompanied by results considered detrimental.

Figure 18 shows the global circulation of oceanic surface currents. Confusing at first glance[*], the map reveals, with a little study, some simple general patterns, and with still more study, an obvious correlation with atmospheric circulation. Consider the Pacific Ocean in the lower center of the map. Two currents flow westward along the equator (winter in the N hemisphere), one a few degrees N of the equator, the other on or slightly S of it. Between them is a narrow E flowing counter-current. These W-flowing currents are driven by NE and SE Trades, which blow toward, converge at, and rise at the equator. The counter-current provides one way in which the water moving west returns, for it cannot continuously move west and pile up against Asia, leaving the E

---

[*] The map is a polar projection -- a view downward on the S pole, so that E is to the right in the upper half, and to the left in the lower half.

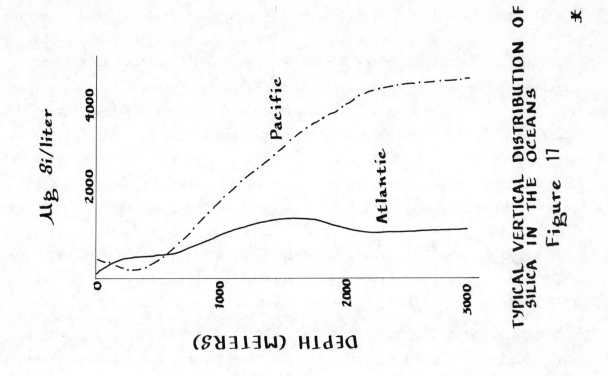

TYPICAL VERTICAL DISTRIBUTION OF
SILICA IN THE OCEANS
Figure 17

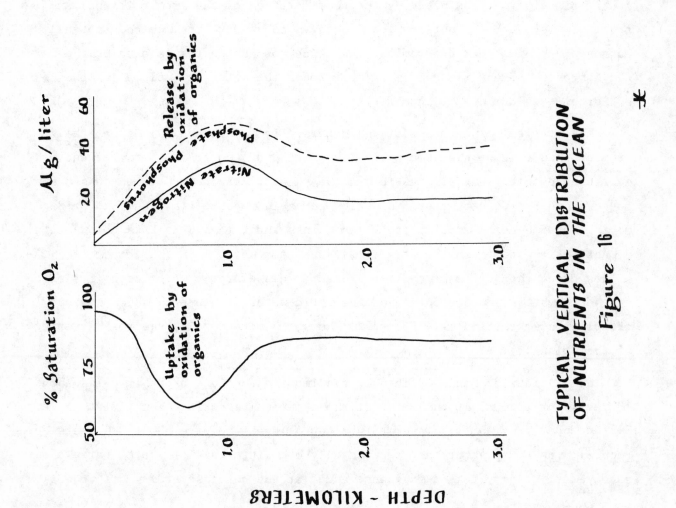

TYPICAL VERTICAL DISTRIBUTION
OF NUTRIENTS IN THE OCEAN
Figure 16

Pacific dry. When the equatorial currents encounter land to the W, they tend to be diverted N and S. One of these 'rivers of the sea' is the Kuroshio or Japan Current, diverted by the Japanese islands and Sakhaline to the north. The current then swings easterly at about 40°N, now driven E by the Wester-lies. It crosses the Pacific, eventually running into N America, where it splits, part going N as the Alaska current, part moving S as the California current, eventually to rejoin the N equatorial current. As the N equatorial current moves, it tends to send off subsidiary currents to the right of its motion. These offshoots from the N Equatorial current and the Kuroshio tend to run into each other, setting up areas at about 20-30°N latitude of rotating surface gyres. The same general pattern is seen south of the equator, except that the development of gyres is even more pronounced. Perhaps the most striking development of a gyre is off the W coast of S America.

The same general circulation pattern is seen in the Atlantic, with the Gulf Stream taking the place of the Kuroshio. The currents that split off the Kuroshio and the Gulf Stream to the N when they strike land result in a complex arctic circulation, dominated by the configuration of the Arctic islands and archipelagos, but in general a loop to the N and W develops, which eventually causes a S-moving current from high latitudes on the eastern shores of the lands. The Labrador current is an example; it keeps the water icy cold along the coast of Maine, even during the summer.

The vertical circulation of the oceans (Figures 19 and 20) is basically simpler than that of the atmosphere, consisting of two great loops between equator and pole. Water is heated at the equator, expands, becomes less dense, moves N and S. Colder water from below moves in to take its place. N or S moving water becomes cold at the Poles, sinks and slides S or N as cold Arctic or Antarctic water. In the Atlantic, dense cold water from the Antarctic slides along the bottom beyond the equator; dense Arctic water moves S along the bottom but slides up on top of Antarctic water. Figure 19 shows oceanic circulation schematically; Figure 20 shows, in more detail, the real layering of the Atlantic.

In considering man's influence, the important points are: it takes about 10 years for renewal of the water of the homogeneous surface zone (mixed layer) by the underlying water; intermediate depth layers require about 500 years for renewal; and deep waters are renewed only over intervals of two or three thou-sand years. It may take a substance added at the coast several years to work

# GLOBAL CIRCULATION OF OCEANIC SURFACE CURRENTS

Figure 18

ADAPTED FROM MUNK, W.R., 1971, THE CIRCULATION OF THE OCEANS, IN OCEANOGRAPHY, READINGS FROM SCIENTIFIC AMERICAN, W. H. FREEMAN, SAN FRANCISCO, P. 65.

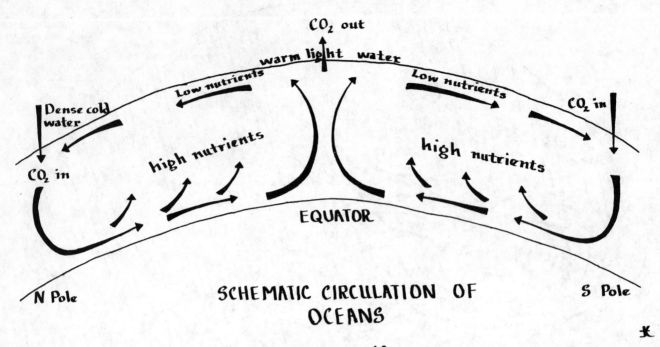

CO₂ out

warm light water

Low nutrients          Low nutrients          CO₂ in

Dense cold water

CO₂ in          high nutrients          high nutrients

EQUATOR

N Pole          S Pole

SCHEMATIC CIRCULATION OF
OCEANS

Figure 19

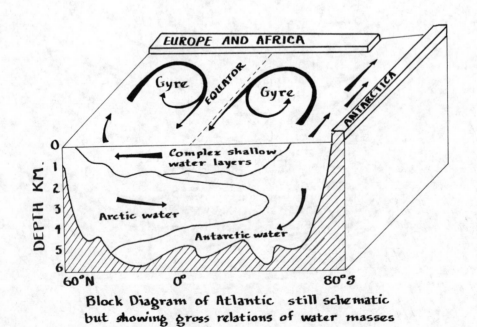

EUROPE AND AFRICA

Gyre          EQUATOR          Gyre          ANTARCTICA

DEPTH KM.

Complex shallow water layers

Arctic water

Antarctic water

60°N          0°          80°S

Block Diagram of Atlantic still schematic
but showing gross relations of water masses

Figure 20

its way across the continental shelf where the shelf is wide, as off the E coast of the U.S., much shorter times on the W coast, where deep water comes close to shore.

In the broad ocean picture, gases such as $O_2$ and $CO_2$ dissolve into the cold Arctic and Antarctic waters, and are released in the low latitudes where waters are warmed as they rise. Thus for any short term (decades) planning, only the upper two hundred meters or so of the oceans can be considered as a sink for a _dissolved_ substance added to shallow water. Dissolved materials added to very deep water will remain deep for long periods, and slowly become dispersed throughout the water mass, unless, perhaps, they are heat producers, like some atomic wastes. In that case, in high concentrations, they could lower water density and cause upward movement. The average velocity of deep currents is 0 to several cm/second; but they may move much faster locally.

Air-Sea Interaction

It should be clear that both atmospheric and oceanic circulation are driven by low latitude heating and high latitude cooling and that ocean surface circulation, of greatest immediate concern, follows a pattern in which wind-generated surface currents are made up of major currents driven by the Trades westward and by the Westerlies eastward, modified by collision with continents, where the currents split N and S. Currents do not move directly down wind, but deflect to the right of the wind direction in N Hemisphere, and vice versa in the S Hemisphere (Eckman spiral), so that the Pacific N Equatorial current, for example, is continuously sending water to the N, and the Kuroshio is shedding water to the S, creating the slowly spinning gyres between the two currents. Gyres are the loci of accumulation of floating material, such as oil and plastic. Speeds of major surface currents range up to several km/hr; the "Kon Tiki" drifted faster than any man can swim in still water (even Spitz!).

Note that the equator is a barrier to exchange of oceanic pollutants between hemispheres, just as it is for the atmosphere.

Breaking waves eject droplets of seawater into the air, where they evaporate, releasing salt particles (aerosols) to the atmosphere. Aerosol composition is similar, but not identical to seawater. Most of the particles fall back on the sea, but some remain in the air and become important constituents of rain on the continents. The droplets also concentrate organic materials which are enriched at the sea surface.

Oceanic circulation is wind and density driven, and both variables are strongly related to the atmosphere. The Trades are initially cool winds, but warm as they flow toward the equator. They pick up moisture from the ocean, increasing ocean surface salinity and causing the water to tend to sink. When they reach the equator they rise and cause clouds and precipitation, thus lowering salinity, and also, because of cloud cover, decreasing solar heating at the equator. Figure 21 shows latitudinal variation of precipitation and evaporation related to these effects.

## Restricted Marginal Basins

In addition to circulating through the atmosphere, pollutants and natural materials are carried into the oceans by streams. The areas of greatest susceptibility are the shorelines and particularly gulfs, bays, seas, and estuaries adjacent to the open ocean. Two major types of circulation develop between the low density stream water and the higher density ocean water, and are critical in determining the fate of materials carried in. The two types can be classified as estuarine and lagoonal circulation. In an estuary, typified by the drowned river mouths of the E coast of the U.S. and the W coast of Europe, fresh water from rivers rides out over the sea water, producing a circulation like that shown in Figure 22. As the fresh water flows out, sea water comes in, driven in part by the tides. Rapid mixing takes place at the interface, there is much biologic activity and a diversity of species adapted to various levels of salt water-fresh water salinities is found. Suspended stream materials are deposited in estuaries, organic materials are oxidized and decomposed, with release of a variety of gases -- $CO_2$, $CO$, $CH_4$, $H_2S$, etc. Oxygen may be much reduced. Metals released from industrial plants upstream of an estuary may come to rest in the sediments of the estuary. Because of circulation and mixing, dissolved pollutional substances may stay in the estuary for a long time as they oscillate up and down.

As stream water mixes with salt water, the suspended particles flocculate and settle much more rapidly than in fresh water, so they are dumped in the estuary, instead of being carried out to sea. As the particles settle they

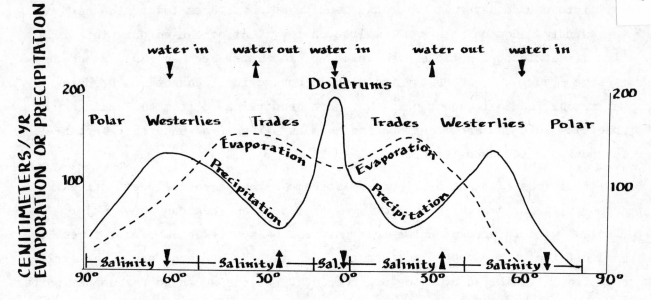

**CENITMETERS/YR**
**EVAPORATION OR PRECIPITATION**

water in   water out   water in        water out       water in

Doldrums

Polar   Westerlies   Trades        Trades   Westerlies   Polar

Evaporation
Precipitation

Evaporation
Precipitation

Salinity   Salinity   Sal.   Salinity   Salinity

90°   60°   30°   0°   30°   60°   90°

Latitude variation of precipitation and evaporation
and gross relation to wind belts

Figure 21

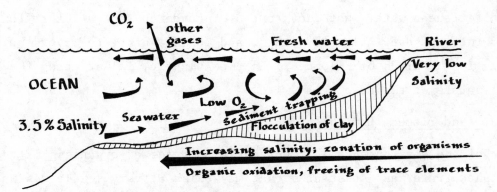

$CO_2$   other gases        Fresh water        River

OCEAN        Very low Salinity

Low $O_2$   sediment trapping
Flocculation of clay

3.5% Salinity   Seawater

Increasing salinity; zonation of organisms

Organic oxidation, freeing of trace elements

ESTUARINE CIRCULATION. Degree of mixing may range
from well defined fresh and seawater layers to nearly
uniform gradient of salinity from fresh water to oceanic.

Figure 22

fall into the denser salt water and tend to be moved back toward the head of the estuary. Commonly they accumulate on the floor of the estuary in a relatively restricted zone. As they build up, the water becomes shallow and extensive dredging is necessary to maintain navigational channels. Dredging may remobilize toxic substances that have been released from the sediment particles into the water surrounding them as a result of anaerobic bacterial action after deposition.

Additions of plant nutrients by streams may be especially effective in promoting plant growth in shallow arms of estuaries, and can cause major changes in the plant and animal communities. Also, it becomes many times more difficult to clean up pollutants discharged in stagnant areas where circulation is inhibited by dense plant growth. Wastes added to the deeper parts of estuaries may contaminate the whole estuary as they circulate.

In lagoonal circulation (Figures 23 and 24), typified by the Mediterranean (which is also the largest example), fresh water inflow is small, and sea water enters as a surface flow. In the latitudes where evaporation exceeds precipitation (see Fig. 21), surface waters become denser, sink, and flow out as an under-current. In the Mediterranean, flow in and out is restricted to the Strait of Gibraltar. Incoming sea water has a normal salinity of about 3.5%; evaporation increases this to 3.7%. Pollutants dissolved in the Mediterranean tend to be carried down and out. The highly saline but fairly warm deep Mediterranean water that flows out as an undercurrent at Gibraltar slides downward into the deep Atlantic waters, and forms a discrete layer that spreads laterally for hundreds of kilometers in the Atlantic, illustrating the 'stratospheric' nature of the deep ocean.

## Food Chain and Element Enrichment

One of the environmentally important properties of organisms is their ability to concentrate many substances that may exist at very low levels in their environment. This concentration, in some cases, is amplified in the food chain. A fish may graze on algae, be eaten by a lobster, who in turn is captured by an octopus, who ends up on the dinner table. At each link in the chain there may be either selection or rejection of various elements, but the possibility always exists of only successive concentrations, resulting in toxic levels of elements hardly detectable in seawater. The case of mercury is one that will be discussed in detail later. Table 6 gives the concentration factors over

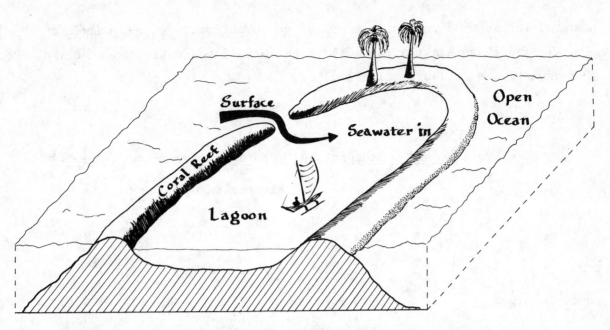

## CLASSICAL LAGOONAL CIRCULATION

Water enters at high tide, drains at low tide. Surface evaporation causes underflow out, heavy rains low salinity surface flow out.

Figure 23

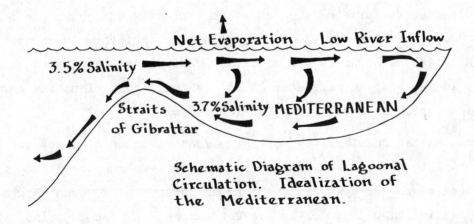

Schematic Diagram of Lagoonal Circulation. Idealization of the Mediterranean.

Figure 24

seawater for some of the basic foods of sea creatures, algae, seaweeds in general, and phytoplankton, as well as the actual concentration achieved. (The numbers are not very reliable!)

Table 6

Concentrations and Concentration Factors of Some Trace Elements*

Element in ppb

| | Cu | Ni | Pb | Co | Zn | Mn |
|---|---|---|---|---|---|---|
| conc. in seawater | 2 | 7 | 0.03 | 0.05 | 2 | 2 |
| conc. in algae | – | 14000 to 80000 | | | 1800 | 2000 to 60000 |
| conc. factor | | 2000 to 4000 | | | ~900 | 10000 to 30000 |
| conc. in seaweeds | | 3850 | | | 1800 | |
| conc. factor | | 550 | | | 900 | |
| conc. in plankton | 800 to 180000 | 140 to 56000 | 0.9 to 360 | | | |

*from many sources

Concentration factors in animals can be extremely variable; as an extreme, the body fluids of tunicates (descendants to primitive vertebrates) may contain thousands of parts per million vanadium, yet the concentration of vanadium in seawater is estimated at only about 2 ppb. Thus the natural concentration factor is about a million-fold.

In summary, the possibilities of enrichment of trace elements in the oceans demonstrate that increase of a trace element in surface waters from a normal level of ppb to ppm may have serious consequences, even though it is hard to realize that 1 ppm of an element can be a threat to animal physiology. Nor should we lose sight of the fact that high element concentrations

47

in foods are commonly used beneficially -- i.e., calcium in milk, iron in
prune juice, iodine in kelp.  In addition, trace element deficiencies can
be deleterious to health, just as excesses can.  Finally, when trace elements
are examined in some detail in a later chapter, we will find a terribly complex
situation in which simultaneous excesses of two elements can cause cancellation
or enhancement of the effects that either would have independently.

48

## Study Questions

1. What are the chief conservative substances in seawater in order of decreasing abundance?

2. Hot brines (60°C) have been found in depressions at the bottom of the Red Sea. What can be said of their salinity relative to average seawater?

3. What are the three major limiting nutrients in the ocean?

4. Why does the decline in nutrients that begins at about 800 meters correspond with the oxygen minimum?

5. If nutrients are continuously lost from surface water by sinking of dead organisms, how is the supply maintained?

6. Would you predict that the areas of gyres have high or low productivity?

7. What are the driving forces for equatorial currents and the currents like the Gulf Stream and Kuroshio?

8. Why is there so much more concern about pollution of ocean surface water than of deep water?

9. Would you predict that the salinity of seawater in Hawaii is higher or lower than average seawater?

10. It is claimed by some that a completely stratified ocean would eventually become devoid of life. Why?

11. The average oceanic rate of upwelling of deep water into surface water is estimated at 2 meters/yr. Average N content is about 20 µg/liter. Streams, on the other hand, discharge about $0.35 \times 10^{20}$ g of water to the oceans, with an average $NO_3$ content of 1 ppm. What is the ratio of upwelling supply of N to river supply?

12. Dissolved materials mix throughout a water mass, but do not settle; suspended solids settle to the bottom. If you wanted to minimize the effect of a soluble poison you had to add to the Mediterranean Sea at Gibraltar, where would you place it and why?

13. The very finest mud particles or air-borne dust settle at the rate of about 500 cm/yr. How long will it take such a particle to reach the bottom of the ocean?

14. Is there an equatorial barrier to dispersal of surface pollutants in the ocean?

15. Does the same relation hold for the deep waters?

16. If you put a note into a bottle and put the bottle into the Gulf Stream off S Florida, in the hope of having it reach Great Britain in the minimum possible time, which would be the preferred release point -- the eastern or western side of the Gulf Stream?

17. The rate and nature of dispersal of pollutants is obviously related to the velocity and turbulence of the transporting medium. What is the approximate ratio of dispersal rate in the atmosphere to that in the oceans?

18. How does question 17 relate to the predictability of behavior of pollutants in the atmosphere versus that in the sea?

19. What are optimum conditions for creation of anoxic conditions with estuarine or fjord type circulation?

# NOTES

Chapter 5

THE GLOBAL CYCLE OF WATER AND OTHER MATERIALS – THE EXOGENIC CYCLE

## The Water Cycle

In nature, water, gases and solids are continuously transferred from ocean to atmosphere and vice versa, from land to atmosphere and the reverse, from land surface to underground waters via soils, from soils into streams and from streams to the oceans. Because so many substances are coupled with water circulation, the water cycle will be examined first.

Total water in the earth's surface environment is about $17000 \times 10^{20}$ g, distributed as shown below.

### Table 7

### The Hydrosphere

| | | |
|---|---|---|
| Oceans | $13,700 \times 10^{20}$ g | 80% |
| Pore water in rocks | $3,200 \times 10^{20}$ g | 18.8% |
| Ice | $165 \times 10^{20}$ g | 1.2% |
| Lakes and rivers | $0.34 \times 10^{20}$ g | 0.002% |
| Atmosphere | $0.105 \times 10^{20}$ g | 0.0006% |

In the cycle, water is evaporated from the oceans (and some from the land). Most of it falls back on the oceans, but that part that falls on the land gives us our supply for rivers, lakes, and potable ground water.

The energy budget for solar radiation, which drives the water cycle, is given in Table 8.

About $3 \times 10^{20}$ kcal/yr of solar energy are used to evaporate water (chiefly from the oceans). Total U.S. energy production is about $1.5 \times 10^{16}$ kcal/yr; the ratio is 20,000 to one! We have not yet begun to rival the energy involved

Table 8

Solar Energy Budget

(kilocalories/yr)

| | |
|---|---|
| Total energy from sun to earth | $1.3 \times 10^{21}$ |
| 30% reflected | $3.9 \times 10^{20}$ |
| 47% to heating atmosphere and earth's surface | $6.1 \times 10^{20}$ |
| 23% used in evaporation | $3.0 \times 10^{20}$ |
| 0.2% used in generating winds, waves and currents | $2.6 \times 10^{18}$ |
| 0.0023% used in photosynthesis | $3 \times 10^{17}$ |

in the circulation of water to the atmosphere. Note that water delivery to the land could be grossly changed, however, by changes in reflection (resulting from larger ice caps, more clouds, perhaps more dust), which is an important percentage of radiant energy.

Figure 25 summarizes evaporation-precipitation for land and sea. To get an idea of the physical significance of $10^{20}$ g as a unit of mass, one can visualize a giant ice cube 50 km on each edge. About 1 meter of water evaporates from ocean surface each year (most restored directly to ocean by rain); about 2/3 meter evaporates from land. About 1/3 precipitation received by land runs off in streams. North America, Europe, and Asia all average about the same amount of precipitation and percent of runoff; South America is wetter than they are; Africa is dryer; Australia is much dryer.

The land areas can be divided into those where precipitation exceeds evaporation and those with the reverse relation. The former rarely have fresh water problems; the latter almost always do. In areas of excess evaporation, stream flow is low, groundwater (not renewable in less than centuries) is used up; irrigation is widely practiced. The effect of irrigation is to 'consume' water by increasing evaporation, thereby increasing the salt content of the remaining water. An example of this is the Colorado River (with about 1% of the flow of the Mississippi River): by the time it reaches Mexico its salt content is at the upper limit (1000 ppm) for human, animal, or agri-

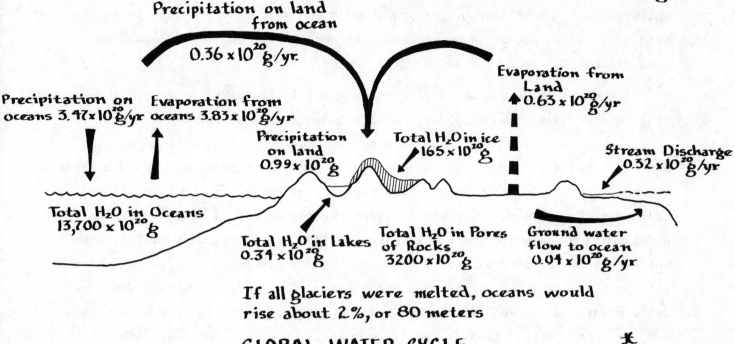

Total H₂O in atmosphere $0.105 \times 10^{20}$ g

Precipitation on land
from ocean
$0.36 \times 10^{20}$ g/yr.

Precipitation on
oceans $3.47 \times 10^{20}$ g/yr

Evaporation from
oceans $3.83 \times 10^{20}$ g/yr

Evaporation from
Land
$0.63 \times 10^{20}$ g/yr

Precipitation
on land
$0.99 \times 10^{20}$ g

Total H₂O in ice
$165 \times 10^{20}$ g

Stream Discharge
$0.32 \times 10^{20}$ g/yr

Total H₂O in Oceans
$13,700 \times 10^{20}$ g

Total H₂O in Lakes
$0.34 \times 10^{20}$ g

Total H₂O in Pores
of Rocks
$3200 \times 10^{20}$ g

Ground water
flow to ocean
$0.04 \times 10^{20}$ g/yr

If all glaciers were melted, oceans would
rise about 2%, or 80 meters

## GLOBAL WATER CYCLE
### Figure 25

ADAPTED FROM GARRELS, R.M., AND MACKENZIE, F.T., 1971, EVOLUTION
OF SEDIMENTARY ROCKS. W.W. NORTON AND CO., NEW YORK.

## SOIL WATER MIGRATION

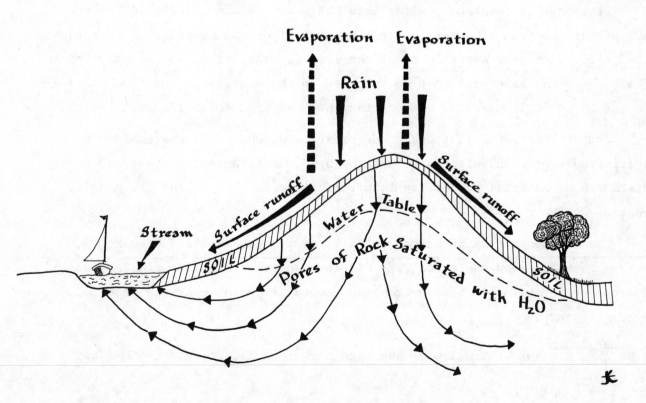

Evaporation    Evaporation

Rain

Surface runoff

Surface runoff

Stream

SOIL

Water Table

Pores of Rock Saturated with H₂O

SOIL

Figure 26

cultural uses.  Most fresh water problems in the next 20 years will be in
arid or semiarid areas.  Many of these are next to the sea where desalina-
tion of seawater (which costs energy!) may help.

Figure 26 shows a general picture of the relation of rain, soil, groundwater and streams.  About 1/3 of the rain percolates down through the soil and
rock.  The upper part of the soil and rock is moist, although the pores are
not saturated with water.  But at some depth (about 6 meters as a world average)
the pores are filled with water.  The boundary of these two zones is the water
table.  At low stream flow there is little suspended material in the stream,
and a maximum concentration of dissolved solids derived by the percolation of
water through the rock and soil.  Most of the dissolved material comes from the
soil.  When it rains there is much surface runoff and soil is washed down the
surface into the stream.  Dissolved solid concentration in the stream diminishes,
but the total carried remains nearly constant.  On steep slopes soils are washed off faster than they can form, leaving bare rock.  Soils contain organic debris from plants and animals which, in well drained areas, is subjected to oxidation.  Dissolved solid composition in streams depends on the nature of the
rock in the drainage area, and thus has tremendous local variations.  But on
the continental scale it averages out to the world river water composition given
in Table 5.  Minor element concentrations are also given in Table 5.

Discharge of suspended solids into the ocean is about $180 \times 10^{14}$ g/yr;
80% of this is from S.E. Asia.  The discharge of dissolved solids is about $40 \times
10^{14}$ g/yr.  Each year about 2 $km^3$ of the land is transported to the sea, and on
the average the same mass must be returned to the continents, or they would disappear in 60 million years, whereas they have endured for billions.

Judson[*] estimates that due to deforestation, change from grassland to
cultivated crops, and effects of grazing by domestic animals, the present
discharge of material to the sea is more than twice that of the pre-man discharge.

The residence time of stream water is short;  at most a few weeks in the
stream channel.  The residence time of groundwater is highly variable, but the
average age of water withdrawn from the ground is probably of the order of hundreds of years.

---

[*] Judson, S., 1968.  Erosion of the land, American Scientist 56, 356-374.

## Lakes

Lakes have been particularly susceptible to change as a result of man's activities.  In the temperate zone, lakes tend to stratify in the summer, with warm light water on top, and dense cold water below.  If they are deep enough, the bottom waters are not stirred by wind and waves, and thus tend to become oxygen-depleted.  In the autumn, surface waters cool and sink; the lake becomes mixed and re-aerated.  If the surface waters cool below 4°C, their density diminishes;  so if ice forms, the water immediately below it is less dense than the deeper waters and density circulation ceases.  Ice prevents wind-stirring, so in winter the lake may again stop circulating, with denser but slightly warmer water at depth rather than at the surface.  Again, oxygen in deep water tends to be used up, and reduced gases, produced by decay, may accumulate in the bottom water.  In spring the ice melts, the surface waters warm through the maximum density point at 4°C, then they sink and the lake 'turns over' again. There are many variations of this classic cycle of biannual renewal:  in the tropics deep lakes may remain stratified the year round.  The Great Lakes are so large that the circulation caused by the cooling and sinking of maximum density surface water, which is replaced by somewhat warmer deep water, is not sufficient to cool the whole body of water to maximum density, and hence they never freeze over completely.  Stratification in these lakes takes place only once a year, in the summer season.

During stratification, if there is enough organic material in the deep water, oxygen can disappear completely, resulting in changes in the bottom fauna and allowing production of gases such as $H_2S$ and $CH_4$.  Shallow lakes are kept stirred by wind and waves so they can survive summer stratification;  but lakes of intermediate depth are most susceptible to oxygen deficiency (e.g., Lake Erie, which is less than 1/5 as deep as the other Great Lakes).  Addition of excess plant nutrients promotes growth of plants, and aids the deoxygenation process in small lakes by reducing wave action and mixing.  Even rivers, which ordinarily are well mixed and oxygenated, can be overwhelmed by excess organic material.

## The Material (Exogenic) Cycle

### Introduction

About 80% of the present day continents are covered with sedimentary rocks. (See Appendix I, p.179).  Of these, about 60% are shales, composed of clay minerals (compositionally complex Al-silicates) and quartz ($SiO_2$); 20% are carbonate rocks, including limestones ($CaCO_3$) and dolomites ($CaMg(CO_3)_2$); 15% are sandstones (largely quartz); and about 5% are evaporite deposits, layers of gypsum ($CaCO_4 \cdot 2H_2O$), anhydrite ($CaSO_4$), and salt (halite, NaCl). The average composition of sedimentary rocks is given in Table 9. The sedimentary rocks are an irregular carpet overlying crystalline rocks. 'Crystalline rocks' is a blanket term covering a wide variety of igneous rocks and sedimentary rocks that have been heated and recrystallized. The crystalline rocks are made up of silicate minerals, dominated by feldspars (K, Al, and Ca-Al silicates) and quartz, with subsidiary silicates of Ca, Mg and Fe. The average composition of crystalline crust is also given in Table 9. Figure 27 shows a schematic cross-section of a continent.

Although the mass of sediments is only about 1/6 that of the continental crust, and the sedimentary layers average only about 5 km thick, as opposed to 35 km for the underlying crystallines, sediments seem to have blanketed most of the crystallines continuously for hundreds of millions of years. Thus today and through the geologic past, the processes of transport of materials from the continents to the oceans to make carbonate rocks, sandstones and shales have been matched by a return of these sediments to positions above sea level.

The ways in which sediments become land again are many and varied. Sometimes the area in which they were laid down was simply uplifted vertically, so that the seas drained away and the sediments remained as nearly horizontal layers. In other instances, as represented by marine sediments of the high Himalayas, the ocean floor was uplifted, broken and folded, as the subcontinent of India in its slow drift northward collided with Eurasia, and a great segment of crust to the south plunged under and raised the Himalayan region.

The general picture of material transport through geologic time seems to be a cannabalistic cyclic one. New sediments are derived from older ones, or by weathering of exposed cyrstalline rocks. New crystalline rocks can be made by the melting of deeply buried sediments and injection of the melt up into the crust, eventually to be exposed to erosion. It is known that the cycle is not entirely 'closed', but to consider it so is a useful first approximation.

Table 9

Average Composition of Earth's Crust, Sedimentary Rocks,
and Suspended Load of Streams

|  | Average Composition Crust (wt %) | Average Composition Sedimentary Rocks (wt %) | Average Composition Suspended Stream Load (wt %) |
|---|---|---|---|
| $SiO_2$ | 63.5 | 59.7 | 61.0 |
| $Al_2O_3$ | 15.9 | 14.6 | 14.3 |
| $Fe_2O_3$ | 2.9 | 3.5 | 9.0 |
| $FeO$ | 3.3 | 2.6 |  |
| $MgO$ | 2.9 | 2.6 | 2.2 |
| $CaO$ | 4.9 | 4.8 | 3.8 |
| $Na_2O$ | 3.3 | 0.9 | 1.4 |
| $K_2O$ | 3.3 | 3.2 | 2.3 |
| $CO_2$ | – | 4.7 | – |
| $H_2O$ (110°C) | – | 3.4 | 5.7 |

(R.M. Garrels and F.T. Mackenzie, 1971; Evolution of Sedimentary Rocks, W.W. Norton & Co., New York, 397 pp.)

------------------------------------------------------------------------

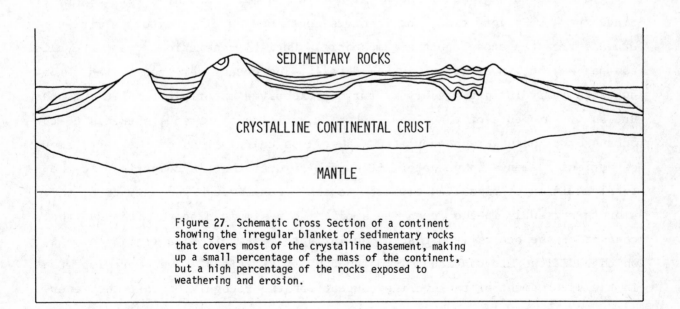

SEDIMENTARY ROCKS

CRYSTALLINE CONTINENTAL CRUST

MANTLE

Figure 27. Schematic Cross Section of a continent showing the irregular blanket of sedimentary rocks that covers most of the crystalline basement, making up a small percentage of the mass of the continent, but a high percentage of the rocks exposed to weathering and erosion.

57

Table 10 summarizes the natural agents that carry materials to the ocean, with some comments on the nature of the materials. Transport of materials is dominated by streams. They carry materials dissolved in the water, and also as solid particles suspended in the water or rolled along the stream bed. Estimated average compositions of the two types of load during geologic time are given in Table 11. It is thought that in the pre-man world the dissolved load was about the same as it is today, and that the suspended load (as given in the Table) was about equal to the dissolved load. As a result of human activities (overgrazing, deforestation, urban development, etc.) the suspended load today is estimated to be 3 times greater than the pre-man average. On the global scale this major but little publicized aspect to man's impact on the earth may be one of the most important long-term 'pollutional' influences.

The solution load of streams, however, is of greater interest to us than the suspended load, when considering most pollutional problems. Many of the dissolved materials are derived by reactions of minerals with atmospheric constituents, and they influence ocean and atmosphere compositions as they enter the ocean and precipitate from it.

The Steady-State System

As has been suggested, the cycling of earth materials has been taking place for a very long time, and involves continuous additions to and withdrawals from the ocean and the atmosphere. Until a generation ago, the cycling concept was not investigated in any detail. The oceans were considered to be an essentially infinite reservoir for the dissolved materials carried into them. But as measurements of the global rates of addition of various elements to the oceans by streams were accumulated, as well as calculations of the total amounts of various elements in seawater, it became evident that in comparison to the total addition of dissolved elements to the ocean through geologic time, the amounts currently in the ocean are small. This simple discovery changed the concept of the oceans from that of an infinite reservoir to a finite one, in which additions are continuously matched by subtractions. Even more important in the development of the cycling concept was the recognition that the reservoirs of elements in sedimentary rocks, although much larger than those in the oceans, are still not large enough to have been continuously weathered and eroded through geologic time without having been renewed 5 or 6 times.

Table 10

Agents of Material Transport To Oceans

| Agent | % of Total Transport | Remarks |
|-------|---------------------|---------|
| Streams | 89 | Present dissolved load 17%, suspended 72%; during geologic past more nearly equal. |
| Ground water | 2 | Estimate poor, dissolved materials like those of streams. Major area of ignorance with respect to possible contamination. |
| Dust | 0.2 | Dust to ocean related to deserts and wind patterns. Sahara major source for tropical Atlantic. Composition similar to average sedimentary rock; many dusts have high (30%) organic content. |
| Shore erosion | 1 | Silts, muds, and sands eroded from shorelines by waves, tides, and currents. Composition like suspended load of streams. |
| Ice | 7 | Ground up rock debris as well as material up to sizes of boulders. Chiefly from Antarctica and Greenland. Distributed in northern and southern seas by icebergs. Composition similar to average sediments. |
| Volcanic | 0.3 (?) | Lavas and gases transported from earth's interior. Amounts and compositions of gases poorly known, but include $CO_2$, $CH_4$, $H_2S$, $SO_2$, $NH_3$, $H_2$. Dusts from explosive volcanoes may be important in climatic control. No one knows how much material from volcanoes is new to exogenic cycle. |

Adapted from R.M. Garrels and F.T. Mackenzie, 1971. Evolution of Sedimentary Rocks, W.W. Norton, New York, pp. 112-113.

59

Table 11

Natural Cycles of the Major Elements

Geologic Stream Fluxes

(units of $10^{14}$ moles/yr)

| Element | Dissolved Load[1] | Suspended Load |
|---------|-------------------|----------------|
| Na | 0.0365 | 0.0156 |
| K | 0.0184 | 0.0148 |
| Mg | 0.0550 | 0.0116 |
| Ca | 0.1260 | 0.0152 |
| Si | 0.0900 | 0.3985 |
| Al | ----- | 0.1114 |
| Fe | ----- | 0.0357 |
| S | 0.0189 | 0.0015 |
| Cl | 0.0270 | ------ |
| $C_{inorg.}$ | 0.3729[2] | 0.0166 |
| $C_{org.}$ | ----- | 0.0583 |

[1]Corrected for atmospheric cycling

[2]Includes C from atmospheric $CO_2$ used in rock weathering.

Adapted from R.M. Garrels and E.A. Perry, Jr., Cycling of Carbon, Sulfur, and Oxygen Through Geologic Time. The Sea, vol. 5, Wiley Interscience, New York, 1974, pp. 303-336.

Table 12 gives the residence times of some major elements in the oceans and in sedimentary rocks. First the reservoir size was estimated. For sedimentary rocks, the reservoir size for a given element is obtained from the total mass of existing sedimentary rocks and the percentage of the element in the rocks. Organic carbon makes up about 0.5% of sedimentary rocks; the estimated total rock mass is 25,000 x $10^{20}$ g, so 25,000 x $10^{20}$ g rocks x 0.005 = 125 x $10^{20}$ g of carbon, or about 125 x $10^{20}$ g/12g/mole = 10 x $10^{20}$ moles. Similar calculations yield reservoir sizes for other elements in sedimentary rocks and the oceans.

Next, the residence time was calculated by dividing the reservoir size by the estimated annual flux of material into or out of its reservoir. Streams bring Ca to the oceans at the rate of 0.1260 x $10^{14}$ moles/yr (Table 11); total Ca in the oceans is 0.15 x $10^{20}$ moles; therefore the residence time is 0.15 x $10^{20}$ moles/0.1260 x $10^{14}$ moles/yr = 1.2 x $10^{6}$ years. Thus streams add an amount of Ca to the oceans each year sufficient to give the present oceanic Ca in about a million years. If Ca had been fed to the oceans at the same rate through 3 billion or more years, without any removal, present oceanic content would be roughly 3000 times the present amount! Only Na and Cl have long residence times in the oceans -- about 100 million years. But 100 million years is short enough compared to all of geologic time that Na and Cl have cycled through the oceans 30 or more times.

The current concept, clearly an approximation, is that all components of sedimentary rocks have been continuously cycling into and out of the oceans, and that one can put together the total complex system by determining the cycles for individual components of the rocks, then adding the cycles together.

Figure 28 shows some estimates of the reservoir sizes of various components of sedimentary rocks, and interactions of the components with the ocean and atmosphere as they cycle. No attempt will be made to discuss the Figure in detail; later chapters are devoted to the total cycles of individual elements. Instead, the message to be received from the Figure is that many of the major components of sedimentary rocks interact with the atmosphere, and that constancy of atmospheric composition depends on a continuous balance among all parts of the system. The $CaCO_3$ reservoir, at the bottom of the diagram, is representative of several others. The flux of $CO_2$ entering the reservoir from below is the

Table 12

Reservoir Sizes and Residence Times of the Elements
(units of $10^{20}$ moles and $10^6$ years)

| | Oceans | | | Sedimentary Rocks | |
| Element | Reservoir | Residence Time[1] | | Reservoir | Residence Time[3] |
| --- | --- | --- | --- | --- | --- |
| Na | 7.06 | 193 | | 15.00 | 288 |
| K | 0.15 | 8.2 | | 12.97 | 399 |
| Mg | 0.82 | 15 | | 25.42 | 381 |
| Ca | 0.15 | 1.2 | | 49.66 | 351 |
| Si | 0.0014 | 0.016 | | 220.03 | 450 |
| Al | -- | -- | | 56.58 | 504 |
| Fe | -- | -- | | 17.17 | 481 |
| S | 0.42 | 22 | | 4.91 | 242 |
| Cl | 8.24 | 305 | | 5.90 | 218 |
| $C_{inorg.}$ | 0.033 | 0.088[2] | | 50.84 | 381 |
| $C_{org.}$ | -- | -- | | 10.42 | 417 |

| | Atmosphere | | | Biosphere | |
| Element | Reservoir | Residence Time | | Reservoir | Residence Time |
| --- | --- | --- | --- | --- | --- |
| $N_2$ | 1.41 | -- | $C_{org.}$ | 0.0042 | 0.078 |
| $O_2$ | 0.38 | 7 | (living and dead organisms) | | |
| $C_{inorg.}$ | 0.00054 | 0.19 | | | |

[1] Corrected for atmospheric cycling

[2] Includes C input from atmospheric $CO_2$ used in weathering

[3] Includes input flux of both dissolved and suspended stream loads (Table 10)

from: R.M. Garrels and F.T. Mackenzie, 1972. A quantitative model for the sedimentary rock cycle, Marine Chemistry, 1, 27-41.

# THE SEDIMENTARY ROCK CYCLE

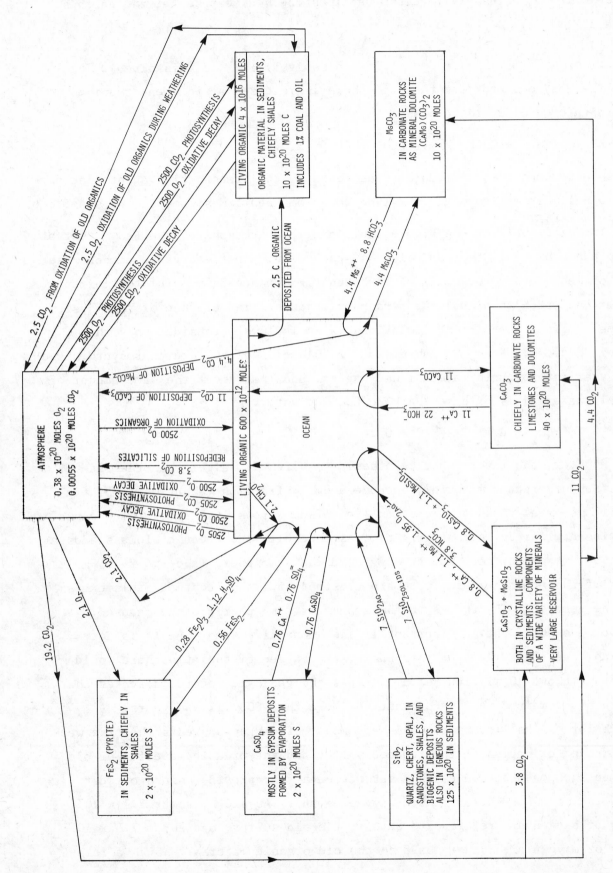

**FIGURE 28**

SCHEMATIC STEADY SYSTEM SUGGESTING AVERAGE RELATIONS OVER THE PAST FEW HUNDREDS OF MILLIONS OF YEARS. NOT ALL IMPORTANT RESERVOIRS ARE SHOWN. TRANSFER OF SOLID MATERIALS FROM VARIOUS RESERVOIRS TO THE OCEAN IS NOT SHOWN; EMPHASIS ON TRANSFER OF DISSOLVED MATERIALS TO OCEAN, AND ON THE REVERSAL OF THE SOLUTION REACTIONS AND RESTORATION OF THE ATMOSPHERE. FLUXES ARE IN UNITS OF $10^{12}$ MOLES/YR.

atmospheric $CO_2$ necessary to dissolve limestone beds exposed at the earth's surface.

$$CaCO_3 + CO_2 + H_2O \rightarrow Ca^{++}_{(dissolved)} + 2HCO_3^{-}_{(dissolved)}.$$

The $CO_2$ consumed from the atmosphere is released again when organisms precipitate $CaCO_3$ in the oceans.

$$Ca^{++} + 2HCO_3^{-} \rightarrow CaCO_3 + CO_2 + H_2O.$$

The $CO_2$ used and released can be considered as the transporting agent that permits old limestones to move from land to become new limestones on the sea floor.

The upper half of the Figure, including the atmosphere, the $FeS_2$ reservoir, the oceanic biomass, the land biomass, the ocean, and the organic carbon reservoir of sedimentary rocks, is of paramount importance in many of the current pollutional problems involving atmospheric gases. The essence of this system of reservoirs and their interrelations is as follows. Consider only the oceanic biomass. Note that about $2500 \times 10^{12}$ moles of $CO_2$ are photosynthesized each year into organic material and oxygen, and that about the same amount of organic material is oxidized back to $CO_2$ and water:

$$CO_2 + H_2O \rightleftarrows CH_2O + O_2.$$

However, the rate of photosynthesis is slightly larger than the rate of oxidation of organic material, because a small fraction (about 0.2%) of the organic material sinks to the ocean floor (see arrows just below the oceanic organic material on left and right). As a result, the oceanic biomass adds a small but important net flux of oxygen to the atmosphere (about $5 \times 10^{12}$ moles/yr). If the atmosphere is to remain constant in its $O_2$ content, this excess must be removed. The change in the amount of atmospheric oxygen caused by this excess would not be important in the human life span, but it is not tolerable geologically. On this basis, the oxygen of the atmosphere would double in about 10 million years, whereas the geologic record indicates that oxygen has remained nearly constant for more than 600 million years. A search for 'sinks' for this small net amount of oxygen suggests that the weathering and oxidation of the reduced sulfur and the organic material of old sediments exposed at the earth's surface each year provide sinks of just about the size needed for annual oxygen demand. As shown, about $2.1 \times 10^{12}$ moles of oxygen are required to oxidize the old sulfur, and about $2.5 \times 10^{12}$ moles of oxygen are used to oxidize the old organic material. Such a rela-

tion suggests that the system has achieved a dynamic balance characterized
by the condition that the amount of oxygen resulting from organic material
lost from the oceanic biomass by sinking is equal to that used   by old
organic carbon and sulfur, and furthermore that the possibility of a change
in this relation might produce a feed-back to restore the balance.

## Feed-back Mechanisms

To comprehend the positive feed-back mechanisms that tend to preserve the
$O_2$ balance, assume that for some reason the oxygen in the atmosphere were suddenly increased. If so, more of the ocean biomass would be oxidized, and less
organic matter would sink to the sea floor, so production of oxygen by this
method would be reduced. On the other hand, oxygen removal by oxidation of sulfur and of old organic material exposed to weathering would remain the same, so
that there would be a net withdrawal of $O_2$ from the atmosphere, and a tendency
for it to return to its original value. Reduction of atmospheric oxygen would
increase the rate of organic material storage in sediments, and would increase
$O_2$ addition to the atmosphere. The amount removed by 'sinks' would either stay
the same or decrease, causing net increase in atmospheric oxygen toward the
original level.

This oversimplification of the feed-back system describes the essence of
the responses of the earth's metabolic system to changes. It should already be
apparent that estimates can also be made of the times required for restoration
of the system. If atmospheric $O_2$ were doubled to $0.76 \times 10^{20}$ moles, the maximum
rate of restoration would be of the order of $2-5 \times 10^{12}$ moles/yr (the maximum that
could be used by the 'sinks', assuming that the rate of organic carbon burial
were nearly zero), so the lowering of $O_2$ to its present level, a removal of 0.38
$\times 10^{20}$ moles, would require at least $0.38 \times 10^{20}/5 \times 10^{12}$ = 7.6 million years.

The value of such information is that it gives us an estimate of response
times to perturbations of the earth system, and of how long man's effects may
endure.

Figure 29 summarizes the cycles of some of the major substances circulating
in the earth system. On the left are the amounts of materials exposed to weathering and erosion each year, in units of $10^{11}$ moles. The total demands of
$O_2$ and $CO_2$ for this weathering are shown as fluxes from the atmosphere to the
land. Man has increased the flow of $O_2$ from the atmosphere to the land by a

# THE EXOGENIC CYCLE~AVERAGE RATES THROUGH GEOLOGIC TIME

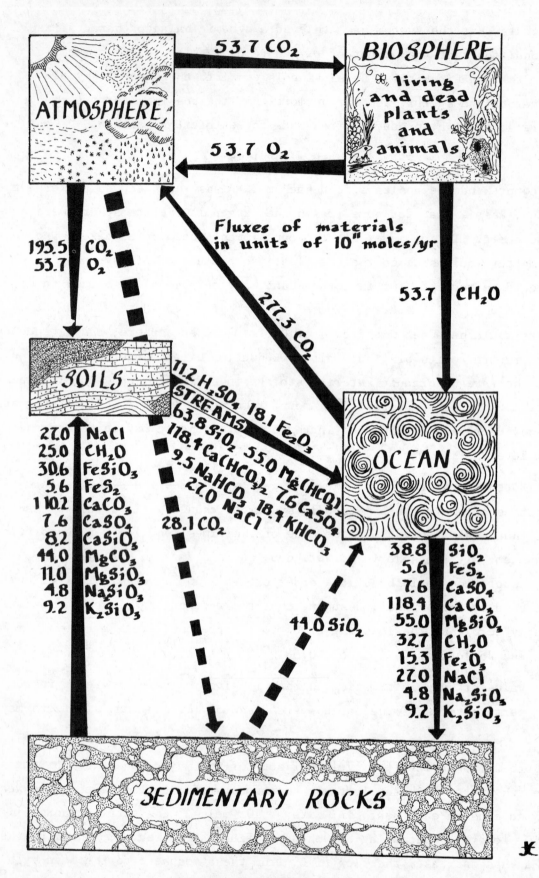

## Figure 29

ADAPTED FROM R.M. GARRELS AND E.A. PERRY, JR., 1973
CYCLING OF CARBON SULFUR AND OXYGEN THROUGH GEOLOGIC
TIME.  THE SEA, VOL. 5, WILEY INTERSCIENCE, N.Y., P. 312.

factor of 60 or more because of the burning of fossil fuels ($C + O_2 = CO_2$).
The flux of $CO_2$ has been reversed; the amount going from land to the atmosphere from fuel burning is about 20 times the amount moving to land from atmosphere by natural weathering.

In the pre-man cycle, the oxygen demands of weathering of the land were restored by photosynthesis, as shown by the flux from the atmosphere to the biosphere. The drains of $CO_2$ from the atmosphere as a result of weathering of the land and of photosynthesis were restored by precipitation reactions in the ocean, chiefly the precipitation of carbonates and silicates:

$$Ca^{++} + 2HCO_3^- = CaCO_3\downarrow + CO_2\uparrow + H_2O$$

and

$$Mg^{++}_{(dissolved)} + SiO_{2(dissolved)} + 2HCO_{3\ (dissolved)}^- \rightarrow$$
$$MgSiO_3\downarrow + 2CO_2\uparrow + H_2O.$$

## Summary

In summary, the exogenic cycle, before changes induced by man's activities, can be regarded as a system of many reservoirs, with a complex web of circulation of materials between them. The system apparently remained in dynamic balance for at least several hundreds of millions of years. Circulation from one reservoir to another certainly was not constant, but an initial change in the size of a reservoir as a result of increased or decreased circulation rates into the reservoir tended to be leveled out by relatively slight readjustments in the sizes of other reservoirs. The overall picture is one of continuous fluctuation of reservoir sizes, but with increases and decreases averaging out through time, without ever having a disturbance in the system sufficient to change the ocean and atmosphere reservoirs of $O_2$ and $CO_2$ beyond the limits of existence of most organisms.

## Natural Production of Some Gases

Only the gases $O_2$ and $CO_2$ have been discussed in the framework of the material cycle. Other species of importance in pollutional problems have natural cycles -- nitrogen oxides, methane ($CH_4$), carbon monoxide (CO), ammonia ($NH_3$), hydrogen sulfide ($H_2S$), and many others.

Gases are produced by nature in many ways. Most spectacular are the releases by volcanoes, which emit water vapor, sulfur gases, methane, carbon gases, hydrogen, and nitrogen gases. As emitted, most of these are not associated with oxygen, and species like $H_2S$, $NH_3$, and $H_2$ are common. Their sources are still in doubt, even though it can be proved that in many cases the lava came from 50 or more kilometers beneath the surface. Much of the gas may be from material encountered by lava en route to the surface, such as seawater, or organic materials in rocks. In any case, volcanic addition is a small % of global output of gases.

Most natural production of gases comes from the cycle of photosynthesis, respiration, and decay, both in the ocean and on land. Recent estimates (see Fig. 28) are that terrestrial and oceanic photosynthesis are roughly equal, somewhere on the order of $5000 \times 10^{12}$ moles of carbon fixed and released each year. Carbon, nitrogen, and sulfur, major (pollutional) elements added to the atmosphere by man, occur in terrestrial plants in the ratios of C/N/S = 160/1.8/1 and they cycle between atmosphere and plants and animals roughly in that ratio. Details of the cycles will be examined later. An oversimplified model has the reduced gases of these elements released in the soil, or in shallow ocean waters, by bacterial decomposition. Much of the gas may be oxidized within the soil or water, and never reach the atmosphere in reduced form, or it may be fixed as a non-volatile compound. Under other conditions, as in water logged soils, the reduced gases, or partly oxidized ones, may reach the atmosphere in that form.

The ratios of elements in various types of organic materials are poorly known. For living terrestrial plants we have derived ratios of C/O/N/S/P of about 800/800/9/5/1 by averaging and weighting numerous sources. Soil humus, living marine plants and organic materials in sedimentary rocks are markedly different from land plants, but tend to be similar in composition to each other. For them we estimate ratios of C/O/N/S/P of 106/106/16/2/1. Chief uncertainty is in the S content of all organic types.

Very roughly, then, land plants have C/N/P ratios of 800/10/1, and marine plants have C/N/P ratios of 100/15/1. Land plants have a much higher C/N ratio than do marine plants, and a somewhat lower N/P ratio.

There are many estimates of the composition of the biosphere (which usually includes living terrestrial and marine plants and their dead residues). These estimates range widely, because of differences between the estimates of the average composition of each type of material, and because of differences in the estimates of the mass of each type.

The oxidation sequences of the initial decay products of organic matter are shown <u>schematically</u> below:

$$\xrightarrow{\text{oxidation}}$$

$$2CH_4 + 3O_2 = 2CO + 4H_2O; + O_2 = 2CO_2;$$

$$2NH_3 + 3O_2 = 2N_2 = 6H_2O; + O_2 = 2N_2O; + O_2 = 4NO;$$

$$4NO + 2O_2 = 4NO_2; + O_2 = 2N_2O_5; + H_2O = 2HNO_3;$$

$$2H_2S + 3O_2 = 2SO_2 + 2H_2O; + O_2 = 2SO_3 + 2H_2O = 2H_2SO_4;$$

$$2H_2 + O_2 = 2H_2O.$$

The details of the mechanisms of these oxidations are complex; in many instances the mechanisms are unknown, but broadly we do see oxidation of the reduced decay products in the atmosphere to $CO_2$, $H_2SO_4$, and $HNO_3$.

Reduced gases are found in water-logged soils, like rice paddies, or in the sediments of stagnant lakes. Stagnant lakes contain little oxygen, abundant $CO_2$, and also $H_2S$, $CH_4$, and $H_2$. In paddy soils, measurements showed $N_2$, 75-11%; $O_2$, 2.8-0%; $CO_2$, 2.0-20%; $CH_4$, 17-73%; $H_2$, 0-2.2%. Recently CO has also been found.

In 'normal' soils (i.e., well aerated), or in well mixed lakes, these reduced gases become oxidized within the soil or within the lake. In a sense this tells us that nature probably will take care of these reduced gases pretty efficiently, unless we increase global processes of photosynthesis and decay many-fold.

It is often convenient to talk about an '<u>interference index</u>': <u>the ratio of man's additions or consumptions to natural additions or consumptions</u>. The index can refer to any reservoir; ocean, atmosphere, streams, etc. Table 13 gives some recent estimates of natural production of gases versus production from fossil fuel burning.

The 'natural' numbers in the table are probably subject to change to larger ones as better information becomes available. The estimates for man's additions for the year cited are better, perhaps within 20 or 30%. Carbon monoxide is a fascinating example. A few years ago there was no information on natural sources,

Table 13

Gas Production to Atmosphere

(in units of $10^{11}$ moles/yr)

| Gas | A<br>Natural | B<br>Fossil Fuel<br>Burning | Interference Index<br>B/A x 100 |
|-----|--------------|------------------------------|--------------------------------|
| $CO_2$ | 50,000 | 3,650 (1974) | 7% |
| CO | 1,690 | 230 (1974) | 14% |
| $NH_3$ | 65 | 14 (including $N_2O$,<br>NO, $NO_2$)(1970) | 22% |
| $H_2S$ | 30 | 16 (includes $SO_2$)<br>(1970) | 53% |

and it was thought that all CO was added by man. But when natural sources were investigated, CO emerged with an intermediate interference level index!

The major point that can be deduced from Table 13 is that because fossil fuel burning is becoming a substantial fraction of the natural production of gases, we must find out whether or not the natural cycle can bear the extra load.

Estimated natural additions to the atmosphere of C, N, and S are in the ratios 1760/2.2/1; whereas the ratios in land plants are 160/1.8/1. It looks as if only about 10% of the N and S return to the atmosphere before they are oxidized or converted to non-volatile forms in soil or in surface waters.

Table 14 is a convenient reference for U.S. production (1966 and 1968) of dominant pollutants. Note that 43% of all emissions in 1966 and 42% in 1968 were the result of transportation (mainly automobiles and trucks!).

Table 14

Emission Rates of Dominant Pollutant Types
and Sources for 1966 and 1968

(Millions of Short Tons/yr)

These data account for more than 90% of all emissions in the U.S.

| Year | Pollutant Source | CO | Hydro-carbons | $NO_X$ | $SO_X$ | Particu-lates | Total by Source |
|------|------------------|-----|-----|-----|-----|-----|-----|
| 1968 | Transportation | 63.8 | 16.6 | 8.1 | 0.8 | 1.2 | 90.5 |
| 1966 | | 64.5 | 17.6 | 7.6 | 0.4 | 1.2 | 91.3 |
| 1968 | Stationary Combustion Sources | 1.9 | 0.7 | 10.0 | 24.4 | 8.9 | 45.9 |
| 1966 | | 1.9 | 0.7 | 6.7 | 22.9 | 9.2 | 41.4 |
| 1968 | Industrial Processes | 9.7 | 4.6 | 0.2 | 7.3 | 7.5 | 29.3 |
| 1966 | | 10.7 | 3.5 | 0.2 | 7.2 | 7.6 | 29.2 |
| 1968 | Refuse Incineration | 7.8 | 1.6 | 0.6 | 0.1 | 1.1 | 11.2 |
| 1966 | | 7.6 | 1.5 | 0.5 | 0.1 | 1.0 | 10.7 |
| 1968 | Miscellaneous* | 16.9 | 8.5 | 1.7 | 0.6 | 9.6 | 37.3 |
| 1966 | | 16.9 | 8.2 | 1.7 | 0.6 | 9.6 | 37.2 |
| 1968 | Total by Type | 101.1 | 32.0 | 20.6 | 33.2 | 28.3 | 214.2 |
| 1966 | | 101.6 | 31.5 | 16.7 | 31.2 | 28.6 | 209.8 |

*Includes such sources as forest fires, structural fires, coal refuse, agriculture, organic solvent evaporation, and gasoline marketing.

Data Source:  National Air Pollution Control Administration

from:  R.H. Essenhigh, 1971.  Air Pollution from Combustion Sources, Earth and
       Mineral Sciences, vol. 40, no. 7, p. 52.

Study Questions

1.  As a world average, approximately what fraction of rainfall on land runs off to the ocean in streams?

2.  About how much water evaporates from the oceans yearly?

3.  How much solar energy does this represent?

4.  How many kilocalories per person per day would this represent?

5.  What is the ratio of the energy of evaporation of the ocean to the demand of one person/day for food?

6.  Is the transfer of water from sea to land a significant factor in the transfer of other substances?

7.  What is the composition of these seasalt aerosols?

8.  What are the contributors of materials to the ocean?

9.  Why do streams have their maximum concentration of dissolved solids at low flow?

10. Carbonate minerals [chiefly $CaCO_3$ and $(CaMg)CO_3$] make up about 20-25% of sedimentary rocks, which cover about 80% of the land surface. When these rocks are exposed to weathering, what do they take from the atmosphere, and where does it go?

11. What does the weathering of silicate minerals abstract or add to the atmosphere?

12. In the natural cycle, how is the $CO_2$ taken from the atmosphere restored?

13. What earth surface reactions involve atmospheric oxygen?

14. In the pyrite oxidation reaction, what becomes of the sulfuric acid formed?

15. It is estimated that about $0.025 \times 10^{14}$ moles of organic matter are buried in oceanic sediments each year. From the photosynthetic reaction above, this burial represents $CO_2$ removed from and $O_2$ added to the atmosphere. As a result of this burial, if there were no compensation for $CO_2$ and $O_2$ from other processes, how long before the $CO_2$ in the atmosphere would be exhausted?

16. Photosynthesis would almost entirely cease if the $CO_2$ level dropped to about 1/3 the present level. Will the burial of organic materials stop photosynthesis 15 or 20 thousand years from now?

17. What is a likely candidate for the feed-back reaction required above?

18. What presumably would happen if the rate of deposition of new organic materials suddenly increased?

19. Are there other natural restorative mechanisms that might come into play as well?

20. During the geologic past there were times when layers of gypsum accumulated at a rate much higher than the long term average. Suggest a feedback mechanism that might have prevented depletion of atmospheric oxygen at that time.

21. Give a basic argument suggesting that the atmosphere and ocean can be regarded as a steady-state system for the past several hundred million years.

22. Will the addition by man of materials that have been cycling naturally (such as $CO_2$, CO, N-gases, etc.) upset the steady-state systems?

23. What would be the chemical species of N, S, and C expected in a water-logged soil? In a well-aerated one?

**NOTES**

Chapter 6

THE CARBON CYCLE

Chief constituents of the carbon cycle are methane ($CH_4$), carbon monoxide (CO), carbon dioxide ($CO_2$), and organic matter ($CH_2O$). Figure 30 gives
the essential relations as now known for all the major species. The text can
be followed by reference to the Figure. Starting at the left side, it can be
seen that natural global production of methane is about $145 \times 10^{12}$ moles/yr. It
is produced by decomposition of organic material under reducing conditions --
bacteria do the work, and the reaction that forms $CH_4$ can be regarded as the dis-
proportionation of $CH_2O$:   $2CH_2O = CH_4 + CO_2$. Our knowledge of methane fluxes
is poor:  the ocean may be a small source to the atmosphere, but data are not
available. Methane comes chiefly from flooded land -- rice paddies or other
flooded cultivation, swamps, stagnant lakes -- although real possibilities exist
that it may also come from 'normal' soils in significant quantities.

As shown, the average level of methane in the atmosphere is about 1.6 ppm,
constituting a total of $520 \times 10^{12}$ moles. From the flux given and the reservoir
size ($520 \times 10^{12}/45 \times 10^{12}$ moles/yr), the residence time is 3.6 years.

Methane is apparently not toxic at any natural levels.

After methane enters the atmosphere it is oxidized to carbon monoxide, apparently in a steady state situation, thus there is a flux of $145 \times 10^{12}$ moles/yr
of $CH_4$ into the CO reservoir. Oxidation is by $OH^-$ in the atmosphere. As stated
before, the oxidation of methane to $CO_2$ within lakes that contain much $CH_4$ in
their bottom waters indicates that methane addition to the atmosphere from well-
mixed oxygenated surface waters is not likely.

Turning to carbon monoxide, it is noted first that most of the CO in the
atmosphere comes from oxidation of $CH_4$ in the atmosphere. The remainder comes
chiefly from decomposition of organic materials in soils, and presumably only
a small fraction from decomposition in the ocean. The basic reactions are

$$CH_4 + \frac{3}{2}O_2 = CO + 2H_2O,$$

$$2CH_2O + O_2 = 2CO + 2H_2O.$$

# THE CARBON CYCLE

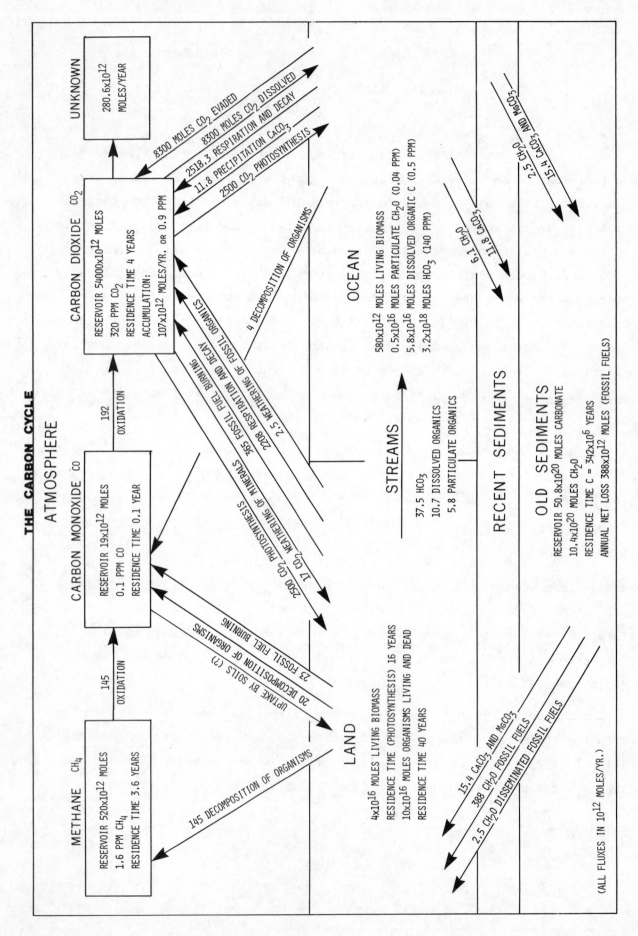

**FIGURE 30**

Natural direct production of CO is highly seasonal, with a minimum in winter in mid-latitude climates, and a maximum in autumn, when falling leaves, etc., release CO from degradation of chlorophyll. The direct flux to the atmosphere is $20 \times 10^{12}$ moles/yr, apparently comparable to current releases from fossil fuel burning (about $23 \times 10^{12}$ moles/yr). However, man's production of CO becomes proportionately greater during winter because of domestic heating increase and decline in natural production.

The level of CO in the atmosphere is about 0.1 ppm CO, corresponding to a reservoir of $19 \times 10^{12}$ moles. The direct flux from the land and from the ocean, plus the indirect flux from $CH_4$ oxidation in the atmosphere, combined with the reservoir size, gives a short residence time of about 0.1 year, or about a month. Oxidation to $CO_2$ takes place according to the basic equation $2CO + O_2 = 2CO_2$. The rate of oxidation is slower in winter than in summer.

The short residence time and the small atmospheric reservoir indicate little likelihood of CO accumulation on a global basis, but it is a serious local problem in urban areas, especially those subject to temperature inversion and pollutant accumulation. In New York City, automobiles (the chief source) produce 3.8 million kilograms of CO each day. Los Angeles has achieved 9.1 million kilograms/day. Concentrations up to 100 ppm are not uncommon in areas of heavy automobile traffic. Wind speed is a critical factor; because of the rapid production of CO and its requirement of days for natural oxidation, dispersion is a much more important removal mechanism than oxidation in local pollution.

Carbon monoxide, like the sulfur gases, is difficult to pin down in terms of the levels that may be harmful. For example, there is a strong correlation between blood CO content of drivers and the number of accidents they have. CO inactivates the oxygen-carrying hemoglobin in the blood. It has been suggested that studies of hundreds of thousands of people in a given area may be necessary to determine the truly toxic levels of CO in terms of exposure time.

Cigarette smokers receive about 400 ppm CO from cigarette smoke; moderate smokers (about 1 pack/day), on the average, have 3.8% of their oxyhemoglobin replaced by carboxyhemoglobin. At 50 ppm CO (the present accepted upper level) about 3% carboxyhemoglobin occurs after very long exposure. Best current estimates are that 5% carboxyhemoglobin increases death rates through a variety of oxygen-related mechanisms -- accentuation of anemia, circulatory failures, etc.

At the 5% level, changes can be measured in driver response times, so high CO in traffic may be an insidious contributor to accidents.

At any rate, the $CH_4$ that goes to the atmosphere, and the CO as well, end up as $CO_2$. The $CO_2$ is photosynthesized to organic matter ($CH_2O$) in plants which provide the organic material at the land surface that gives rise to more CO and $CH_4$. The ocean is apparently a smaller natural source of CO than the land (estimated $4.0 \times 10^{12}$ moles/yr), suggesting again that oxidation in aqueous medium is much more effective than in the atmosphere.

The amount of CO or $CH_4$ carried in solution in streams to the ocean is given as zero. It undoubtedly is small, but not zero, especially in streams with man-induced high values of dissolved or particulate organic matter.

Figure 30 also gives data on the carbon reservoirs. First consider the sizes of the reservoirs of carbon. The atmosphere contains a total of $54,600 \times 10^{12}$ moles of carbon, dominated by $CO_2$. The ratios of $CO_2/CO/CH_4$ are 2850/1/27. The ocean contains $0.0326 \times 10^{20}$ moles of carbon, dominated by dissolved $HCO_3^-$. The ratios of $HCO_3^-$/dissolved C/particulate C are 640/11.6/1. Sedimentary rocks are the largest reservoir of carbon, containing $61.2 \times 10^{20}$ moles, in the ratio carbonate/$CH_2O$ of 4.9/1. The moles of carbon in living and dead organisms in and on the land surface are about $10 \times 10^{16}$.

Thus the ratios of the reservoirs are:

sed rocks/ocean/surface organisms/atmosphere = 110,000/60/2/1.

Some residence times of interest are:

Residence time of C in terrestrial plants as related to photosynthesis:

$$\frac{C \text{ in plants}}{\text{photosynthetic flux}} = \frac{4 \times 10^{16} \text{ moles}}{2500 \times 10^{12} \text{ moles/yr}} = 16 \text{ years.}$$

On the other hand, the residence time in marine plants is only

$$\frac{580 \times 10^{12} \text{ moles}}{2500 \times 10^{12} \text{ moles/yr}} = 0.23 \text{ years.}$$

Whereas the rate of formation of new plants in the ocean and on the land is estimated to be the same, the turnover time in the ocean is much faster.

Because of photosynthesis, an amount of $CO_2$ equal to that in the atmosphere passes through plants in

$$\frac{54,000 \times 10^{12} \text{moles C in atmosphere}}{5000 \times 10^{12} \text{moles C photosynthesized/yr}} = 11 \text{ years.}$$

Also, $CO_2$ exchanges between ocean and atmosphere by solution and evasion show that an amount equal to the atmospheric total goes in and out of the ocean in

$$\frac{\text{moles dissolved C in ocean}}{\text{flux of } CO_2 \text{ into ocean}} = \frac{3.2 \times 10^{18} \text{moles}}{8300 \times 10^{12} \text{moles/yr}} = 385 \text{ years.}$$

The residence time of $CO_2$ in the atmosphere, if all natural fluxes in and out are considered, is only about

$$\frac{54,000 \times 10^{12} \text{moles C}}{13,600 \times 10^{12} \text{moles C/yr}} = 4 \text{ years.}$$

Clearly the atmosphere is delicately balanced with respect to $CO_2$; any significant change in the ratio of photosynthesis to respiration and decay could change $CO_2$ levels several-fold in 100 years.

Fossil fuel burning adds $388 \times 10^{12}$ moles C/yr to the atmosphere (1971), or $388 \times 10^{12}/5000 \times 10^{12}$; or about 8% as much C gases as released by natural respiration and decay. Also, fossil fuel burning accelerates the removal of organic C from the rock reservoir by a factor of $388 \times 10^{12}$ moles/yr/$2.5 \times 10^{12}$ moles/yr = 155. Removal of organic C from the rock reservoir by fossil fuel burning is miniscule compared to the total reservoir of C ($61.2 \times 10^{20}$ moles versus consumption of $388 \times 10^{12}$ moles/yr). The fuels are only that very small percentage sufficiently concentrated to be extracted economically, so that the lifetime of the mineable fossil fuels at current and projected rates is only a few hundred years, with coal furnishing the chief reserve.

In balance, it is found that about $107 \times 10^{12}$ moles $CO_2$/yr are measurably accumulating in the atmosphere, but a much larger amount ($280.6 \times 10^{12}$ moles/yr), which should be accumulating in the atmosphere, has gone somewhere. Its fate is not known; it may go to increasing the $CO_2$ flux into the oceans as dissolved $HCO_3$ in surface water, or into increasing either the oceanic biomass or the terrestrial biomass. At any rate, its amount is so large that to assume it is removed <u>by excess $CH_2O$ burial alone is probably quite inadequate.</u>

It is worth re-emphasizing that the total CO added to the atmosphere through the chain $CH_4$-CO-$CO_2$, or even from CO-$CO_2$ alone, from natural sources, dwarfs that from fossil fuel burning ($169 \times 10^{12}$ moles/yr/$23 \times 10^{12}$ moles/yr). If respiration and decay were doubled (from roughly 5000 x $10^{20}$ moles/yr to 10,000 x $10^{20}$ moles/yr), perhaps by a temperature increase, and photosynthesis remained constant, $CO_2$ in the atmosphere would be doubled in 10 years; if the reverse occurred, and photosynthesis were doubled while respiration and decay remained constant, $CO_2$ in the atmosphere would be halved in 10 years. This suggests that an effective and quick-acting feedback mechanism is at work. In its simplest form, increase in photosynthesis would deplete $CO_2$, which would in turn slow photosynthesis. The magnitude of the ratios of photosynthesis to those of consumption or release of $CO_2$ by natural weathering are so large that it seems unlikely that atmospheric $CO_2$ control, even though there would be weathering response feedbacks, can be related to such processes.

The total increase of $CO_2$ in the atmosphere has been about 15% in the last 100 years; from about 280 to 323 ppm, or about 5074 x $10^{12}$ moles. If four times as much as the increase has been added, then about 20,000 x $10^{12}$ moles have gone elsewhere in the last 100 years. Because the estimates of living and dead surface organics are of the order of 160,000 x $10^{12}$ moles, this would represent an increase of only 13% of the land surface biomass -- well within any estimates that can be made of its actual size. Photosynthetic rate, other factors constant, is proportional to atmospheric $CO_2$ content for several-fold changes, so that the apparent tendency to compensate for increased $CO_2$ could be explained, in essence, by faster-growing plants whose mass and debris are greater than before.

The climatic effects of a higher $CO_2$ level are in doubt. Current modeling of the probable effects (greenhouse) suggests that fossil fuel burning will not raise temperatures more than 2°C, even with high predicted increases in burning rates. Such an increase might have dramatic long-term effects over thousands of years (such as melting ice caps), but the rate should permit easy adaptation by human cultures, such as re-siteing of cities because of sea level rise (200 ft. or so). But we have already experienced ± 2°C variations over periods of tens of years, with only generally beneficial effects, or at least without obviously adverse ones. In fact, global temperatures apparently have fallen 0.3°C for the period 1940 to present.

Study Questions

1. The estimate for total annual photosynthesis is about $5000 \times 10^{12}$ moles of organic material/yr. For the reaction $CO_2 + H_2O = H_2O + O_2$, about 110 kcal of solar energy are required per mole of $CH_2O$ formed. What is the total annual solar energy used in photosynthesis?

2. What % is this of the total solar radiation received at the earth's surface?

3. How much of this solar energy consumed is returned to the atmosphere?

4. Heat flow from the earth's interior is about $4.4 \times 10^{-2}$ kcal/cm$^2$/yr. What is the total heat arriving at the earth's surface/yr?

5. From the above, is heat flow an important item in controlling earth surface conditions?

6. Why is the natural flux of methane to the atmosphere so much greater from land than from the sea?

7. Is the rate of oxidation of methane in the atmosphere slower or faster than that of carbon monoxide?

8. Why are the numbers given for methane and carbon monoxide fluxes to the atmosphere in Fig. 30 probably considerably less than the true values?

9. Bolin (B. Bolin, The Carbon Cycle, Scientific American, 223, 124-132, 1970) estimates the living terrestrial biomass at $3.75 \times 10^{16}$ moles of carbon. What is the residence time of atmospheric $CO_2$ in the living biomass? (Note: moles C = moles $CH_2$).

10. Roughly what % of carbon in organic material photosynthesized on land each year is returned to the atmosphere as methane and carbon monoxide, owing to incomplete oxidation in the soils?

11. If the $280 \times 10^{12}$ moles of $CO_2$ that 'disappear' from the atmosphere go into increasing the living terrestrial biomass, what would be the annual increase in that mass?

12. Would such an increase be detectable?

13. If the well-mixed layer of the ocean is taken as 100 meters thick, and has an average $HCO_3^-$ content of $2.3 \times 10^{-6}$ moles per gram of seawater, what is the total reservoir of $HCO_3^-$ in this layer?

14. Would the annual addition of excess atmospheric $CO_2$ to the oceans be detectable? ($CO_2 + H_2O = H^+ + HCO_3^-$)

15. How about the possibility of measuring directly the increase of the flux of $CO_2$ into the oceans versus that coming out?

16. Why is it more likely that the 'excess' atmospheric $CO_2$, if it is being taken up in the biomass, is going into the terrestrial rather than the marine biomass?

17. Show how the ocean would act as a buffer toward changes in atmospheric $CO_2$.

18. Suggest a mechanism whereby an increase in the terrestrial biomass to a fixed larger size could continuously remove 'excess' $CO_2$ from the atmosphere.

19. Is the increase in atmospheric $CO_2$ (.9 ppm/yr at the moment) a threat to the world's future?

20. If there were a photosynthesis block in the oceans, so that photosynthesis were halted instantly, but if decay and oxidation continued, then

   (a) How long before the oceanic biomass (living only) would decay?
   (b) If all the $CO_2$ went into the atmosphere, what would be the atmospheric content?
   (c) Would there be serious effects on atmospheric oxygen?
   (d) What would happen to N and P in the mixed layer?

   Note: According to Bolin, living oceanic biomass = $0.058 \times 10^{16}$ moles; rate of decay and release of $CO_2$ to the atmosphere = $2518 \times 10^{12}$ moles/yr.

21. If there were a complete photosynthesis block on land and sea, how long for decay and oxidation of all living and dead terrestrial and oceanic biomass, using Bolin's estimates (probably maxima)?

   (Terrestrial biomass living and dead    $10 \times 10^{16}$ moles )
                                                                      } $35 \times 10^{16}$ moles
   Oceanic biomass living and dead    $25 \times 10^{16}$ moles )

22. From #21, how much $CO_2$ would be added to the atmosphere?

23. If this $CO_2$ re-equilibrated with the ocean, what would be the atmospheric content? (See question 17).

Chapter 7


THE SULFUR CYCLE


## Global Cycle

The cycle of sulfur is shown in Figure 31.  Table 15 gives additional docu-
mentation of atmospheric species.  In this instance the major sources and sinks
are shown, rather than the net fluxes.  Considering first the reservoirs, it
is obvious from the diagram that the ocean contains $10^9$ times as much S as the
atmosphere, and, in contrast to most elements in the sea (except Na and Cl), the
sulfur has a substantial oceanic reservoir, even when it is compared to that in
sedimentary rocks.

A major site of action is the land-atmosphere interface.  Weathering of
ancient sediments, containing chiefly gypsum ($CaSO_4 \cdot 2H_2O$) and pyrite ($FeS_2$),
yields substantial amounts of sulfur.  Thus they are sources of both reduced
and oxidized sulfur, in the ratio of roughly 2/3.  Gypsum weathers by simple
solution, so it makes no demands on atmosphere or biosphere, and enters streams
as $Ca^{++}$ and $SO_4^=$ ions.  Pyrite, on the other hand, as shown before, consumes
oxygen to oxidize to $Fe_2O_3$ and $H_2SO_4$:

$$4FeS_2 + 15O_2 + 8H_2O = 2Fe_2O_3 + 8H_2SO_4.$$

It has ordinarily been assumed that the $H_2SO_4$ generated reacts with soil
minerals to neutralize $H_2SO_4$ and produce soluble salts of metal sulfates.  A
typical example could be:

$$CaCO_3 + H_2SO_4 = Ca^{++} + SO_4^= + CO_2 + H_2O.$$

The $CO_2$ generated in the soil would in turn be neutralized by a reaction such as

$$CaCO_3 + CO_2 + H_2O = Ca^{++} + 2HCO_3^-.$$

More recently, the possibility has emerged that oxidation of $FeS_2$ in soils
may not be complete, and that $SO_2$ may be released directly to the atmosphere
by a reaction of the type:

$$2FeS_2 + 5.5O_2 = Fe_2O_3 + 4SO_2.$$

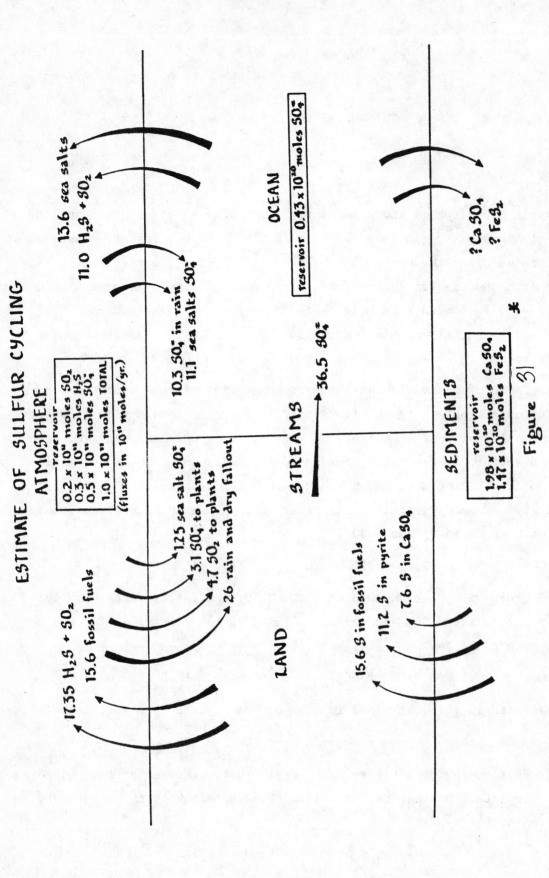

ESTIMATE OF SULFUR CYCLING

Figure 31

Data from many sources: see especially W.W. Kellog, R.D. Cadle, E.R. Allen, A.Z. Lazarus and E.C. Martell, 1972, The Sulfur Cycle. Science, vol. 175, p. 589–596. Also R.M. Garrels and E.A. Perry Jr., 1974, Cycling of Carbon, Sulfur and Oxygen through geologic time. The Sea, vol. 5, Wiley Interscience.

Table 15

Properties of Atmospheric Sulfur Species

Hydrogen sulfide ($H_2S$) Gas, easily soluble in water. Highly toxic. Danger level a few ppm. Residence time a few hours. Source: decay of vegetation, bacterial reduction of sulfate. Natural level 0.2 ppb, or 0.0002 ppm.

Sulfur dioxide ($SO_2$) Gas, easily soluble in water. Highly toxic, acrid smell, eye irritant, causes respiratory ailments. Probable toxic level of exposure: 0.11 ppm, natural level 0.0002 ppm. Source: dominantly fossil fuel burning; annual production $15.6 \times 10^{11}$ moles. Residence time hours or days.

Sulfuric acid ($H_2SO_4$) Gas or particles. Easily soluble in water. Highly toxic, causes respiratory ailments. Strong acid, has important effects in rain. Natural level 0.001 ppm; 0.005 ppm E coast of U.S.; 0.020 ppm, Frankfurt, Germany. Source oxidation of $H_2S$ or $SO_2$. Residence time a few days.

X-sulfate particles (i.e., $(NH_4)_2SO_4$, $CaSO_4$) Source: sea salt or volcanoes. After eruptions, layer forms in stratosphere; residence time 2-3 years. Residence time in troposphere a few days.

---------------------------------------------------------------------

This $SO_2$ would behave in the same way as $SO_2$ released from fossil fuel burning (to plants, to $H_2SO_4$, etc.).

All the sulfur from the weathering of rocks, including that from gypsum and pyrite, is assumed to enter the stream flux as a sulfate, despite the possibility of minor loss of $SO_2$ to the atmosphere from pyrite weathering.

As shown in the diagram, major addition to the atmosphere is $SO_2$ from fossil fuel burning. Coal averages about 20,000 ppm S (2%), oil about 3400 ppm (0.3%), and processed fuels, such as gasoline, even less. Thus coal is the chief contributor to the $SO_2$ additions to the atmosphere. It can be anticipated that coal is destined to be the chief source of energy in the near future, so $SO_2$ additions to the atmosphere will depend on control of coal burning, and upon devices to minimize $SO_2$ release. At the moment the priority is on low S coals, but they are in short supply, and $SO_2$ removal devices (especially because of high toxicity of $SO_2$) have a high priority.

In making a sulfur balance, it is necessary to assume a large flux of $H_2S$ or $SO_2$ into the atmosphere from either sea or land or both. In part this may be $SO_2$ from pyrite weathered in rocks, or it may also be $H_2S$ or $SO_2$ released from water-logged soils, swamps, etc.

The sulfur coming to the land surface from the atmosphere is in minor part from sea salt aerosols, but this is far overbalanced by $SO_2$ that is taken up by plants, or by $H_2SO_4$ or neutralized sulfates (so-called $XSO_4$) that arrive at the surface in rain.

If it is assumed that all the $SO_4$ that arrives on the land in rain is washed into streams, 80% of the stream load would consist of atmospherically derived $SO_4$. But most estimates of the $SO_4$ in streams derived by natural weathering of $CaSO_4$ and $FeS_2$ indicate that 50-60% of the $SO_4$ in streams comes from this source. At the moment, this discrepancy is resolved only by assuming that a high percentage of the $SO_4$ that arrives at the land surface from the atmosphere is reduced in the soil to $H_2S$ or $SO_2$ and returned to the atmosphere before it gets into streams. No clues are available to substantiate this conclusion. Some authorities estimate as much as 30% of river runoff of dissolved sulfur is of man-induced origin (fossil fuel $SO_2$ plus fertilizer $SO_4^=$).

To summarize the situation with respect to sulfur addition and removal from the land, fossil fuel burning (chiefly coal) is an important contributor, but is smaller than natural additions. $H_2S$ is oxidized quickly (hours or a few days); $SO_2$ is oxidized more slowly, but within days, while $H_2SO_4$, the final oxidation product, is chiefly rained out of the atmosphere and arrives at the earth's surface as dilute sulfuric acid or as neutral sulfates (i.e., $(NH_4)_2SO_4$). In some local areas (e.g., New England and Sweden), the increase in $H_2SO_4$ rained out has had important effects on dilute fresh water lakes. Because they are dilute, $H_2SO_4$ makes them much more acid than it would make a saline body of water to which the same amount of $H_2SO_4$ were added, and it can have drastic effects on the biota. Otherwise, the general natural effects are small. The real danger from sulfur gases comes during atmospheric temperature inversions when $SO_2$ and $H_2SO_4$ particles accumulate, and become highly lethal in their effects.

There is no question that the sulfur problem of using coal must be solved.

With respect to the sea, sulfur is released to the atmosphere and returned to the sea in huge quantities, but this is a closed cycle, and not of concern. The sulfur (mostly as $H_2SO_4$ and $XSO_4$) that drifts over the ocean from the land probably is not a problem because it is dissipated in the surface water.

## Summary

Sulfur compounds are dangerous pollutants. Because coal is a major source, and because we must look to more and more coal burning in the next decade or two, it is imperative that procedures be developed for removing sulfur gases during coal combustion. Sulfur pollution is highly toxic locally, where there are temperature inversions and stagnant air masses. On the global scale, sulfur gases are chiefly restricted to the troposphere, although sporadic volcanic eruptions can put $XSO_4$ and $H_2SO_4$ particles into the stratosphere. The return of $H_2SO_4$ to the land, over broad areas hundreds of kilometers downwind from major sources, has been a real factor in changing the ecology of fresh water lakes. Sulfuric acid and particulate iron oxide seeping into rivers from abandoned coal mines and refuse dumps have had bad effects on many rivers, notably the Ohio River system in the U.S. Effects on the ocean are probably minimal.

There seems to be little chance of accumulation of sulfur gases and products in the global atmosphere. Information is needed on the percentage of sulfur received by the land from the atmosphere that goes down streams, the percentage yielded to streams by the natural weathering of rocks, and the amounts coming into streams via groundwater, so that more accurate cycles can be constructed.

## Stable Isotopes as an Aid in Pollutional Studies:
## Sulfur as an Example

Atoms of sulfur occur with several different atomic weights: $^{32}S$ is the most abundant, $^{34}S$ is second, the rest are in trace amounts. The ratio of $^{34}S$ to $^{32}S$ in nature is about 1/22.5, but variations of several percent in the ratio are found in various materials. For convenience, the variations in the ratios are usually expressed as parts per thousand (‰) deviation ($\delta$) from an arbitrarily chosen standard. The standard is sulfur from the Canyon Diablo meteorite, and was chosen because it may well represent the average ratio ($^{34}S/^{32}S$) for all earthly sulfur.

To calculate $\delta^{34}S/^{32}S$, where $\delta^{34}S$ means deviation of the sample ratio from that of the standard, the formula used is:

$$\delta^{34}S‰ = \left( \frac{^{34}S/^{32}S \text{ sample}}{^{34}S/^{32}S \text{ standard}} - 1 \right) 1000.$$

For example, the ratio of $^{34}S/^{32}S$ in seawater is 2% greater than that in the Canyon Diablo meteorite, or 20‰. This relation is usually stated $\delta^{34}S$ seawater = +20‰ (or just +20).

When gypsum precipitates from seawater during evaporation, $\delta^{34}S$ of the gypsum is the same (within 1‰) as that of the seawater. Therefore, measurement of $\delta^{34}S$ of ancient gypsum deposits measures the $\delta^{34}S$ of the seawater from which it came.

On the other hand, when pyrite $(FeS_2)$ is formed in marine sediments, $\delta^{34}S$ of its sulfur averages about 30‰ less than the seawater. This 'fractionation' is caused by the complex series of bacterially controlled reactions that form pyrite by reduction of seawater $SO_4^=$, and partial oxidation of the $H_2S$ formed $(S^=)$ to the sulfur in pyrite $(S^-)$. Interpretations of seawater $\delta^{34}S$ from the $\delta^{34}S$ of the pyrite formed from it are not as simple as those from the $\delta^{34}S$ of gypsum, because, although fractionation averages about 30‰, it is highly variable from one pyrite occurrence to the next.

In the ensuing discussion it will be assumed that the average $\delta^{34}S$ for all gypsum deposits is +18, and the average for all pyrite deposits is -12. Consequently, if the $SO_4^=$ in streams derived from gypsum comes from ancient gypsum deposits of all geologic ages, it should have a $\delta^{34}S$ of +18. Similarly, $SO_4^=$ in streams derived from the oxidation of pyrite, also representing a composite of ages, should have a $\delta^{34}S$ of -12.

Therefore, <u>if</u> the only sources of $SO_4^=$ in stream water were gypsum and pyrite, and if $\delta^{34}S$ of the stream water were known, the relative contributions of gypsum and pyrite to streams could be calculated. The balancing equation is written:

$$(\delta^{34}S \text{ gypsum})(\text{moles gypsum } SO_4^=) + \delta^{34}S \text{ pyrite } (\text{moles pyrite } SO_4^=) =$$

$$(\delta^{34}S \text{ streams})(\text{moles gypsum sulfur + moles pyrite sulfur}).$$

Unfortunately, the global average for $\delta^{34}S$ of streams is not known; the few samples available give values of about +7. Let us use this number to obtain the ratio of gypsum to pyrite contribution to streams in the simplified example that assumes these are the only two sources of sulfur to streams.

$$(+18)(\text{moles gypsum sulfate}) + (-12)(\text{moles pyrite sulfate}) =$$

$$+7(\text{moles gypsum sulfate + moles pyrite sulfate}).$$

18(moles gypsum sulfate)- 12(moles pyrite sulfate) =

7(moles gypsum sulfate) + 7(moles pyrite sulfate)

11(moles gypsum sulfate) = 19(moles pyrite sulfate)

$$\frac{\text{moles gypsum sulfate}}{\text{moles pyrite sulfate}} = \frac{19}{11} = 1.73.$$

This result is inconsistent with the fluxes of S from gypsum and pyrite in Figure 31, which indicates a ratio of 7.6/11.2, or 0.68. The error may be chiefly in the assumption here of $\delta^{34}S$ of +7 for streams.

If the contribution from gypsum and pyrite were equal, $\delta^{34}S$ of streams would have been midway between +18 and -12, or +3.

## Sulfur Cycling During Geologic Time

If, as has been suggested before, the exogenic cycling of various minerals through geologic time has been in a steady state, e.g., if the amount of gypsum carried from the land to the ocean each year were constant and equal to the amount precipitated each year, and the same were true for pyrite, neither the amount of $SO_4$ in the oceans nor its isotopic composition would change. The same mix of isotopes would enter the ocean that came out in the sediments. In this regard, gypsum deposits of various ages up to 100 million years old do give $\delta^{34}S$ values of about +20, that of the present day ocean. A steady-state system for that time interval certainly is indicated. But when older gypsum deposits are examined, major variations in $\delta^{34}S$ are discovered. Gypsum deposits of the Permian Period, about 250 million years ago, have a $\delta^{34}S$ of about +10; in Devonian rocks, some 350 million years old, the values are about +25. What could cause such large differences in $\delta^{34}S$ of the gyspum (which recorded $\delta^{34}S$ of the oceans)?

It has been known for more than 100 years that the Permian Period was a time during which an unusual amount of gypsum formed. Would precipitation of an excess of gypsum over the steady-state amount be expected to lower oceanic $\delta^{34}S$? The answer is yes, because the feed to the oceans is a mix of sulfur isotopes from uplifted gypsum and pyrite deposits, and hence the isotopic composition of the feed is always less than that of the gypsum precipitated and of the ocean itself. Therefore removal of more gypsum than is fed in from gypsum sources can be considered to include 'light' sulfur from the pyrite source. As an extreme, if the feed by rivers is +7 and <u>all</u> the sulfur is removed as gyspum, the time would have to come when the $\delta^{34}S$ of the ocean would equal that of the flux into it (+7). So formation of excess gypsum would tend to 'lighten' the oceanic sulfur.

Thus a decrease in oceanic $\delta^{34}S$ implies that there has been a transfer of 'light' sulfur from pyrite into gypsum, and that some of the $SO_4^=$ in streams derived from pyrite does not get reprecipitated as pyrite, but is instead transferred to the gypsum reservoir.

This conclusion has important implications concerning atmospheric oxygen. If pyrite is oxidized (using 15 moles of $O_2$ for every 8 moles of S) during weathering, and the resultant sulfate is not entirely reduced back to pyrite, then there is continuous removal of oxygen from the atmosphere. Modeling of the system is complicated and controversial, but it appears that the excess gypsum formed during the Permian Period would store at least as much oxygen as is now in the atmosphere, and probably 2 or 3 times as much. Yet, despite the fact that the Permian was a time of pronounced biologic change, it is obvious that atmospheric $O_2$ could not have sunk to zero, or even close to it, without disrupting the fossil record much more than is observed.

The obvious conclusion is that there must have been a feed-back mechanism that restored oxygen as it was used up -- perhaps not on a one-to-one basis, but enough to keep its level perhaps no less than half of that today, perhaps even to hold it close to that of the present. One mechanism suggested (by the authors) is based on the fact that excess deposition of gypsum requires calcium as well as oxygen for sulfate. The Ca required far exceeds that in the oceans, so the only source seems to be limestone ($CaCO_3$). Transfer of Ca from limestone to gypsum would release $CO_2$, which if photosynthesized into organic matter and buried would give $O_2$ to the atmosphere ($CO_2 + H_2O = CH_2O + O_2$). Thus excess gypsum deposition would be accompanied by excess burial of organic matter, with a transfer of carbon from limestones to buried organic matter. The required increase in the flux of organic matter from the oceans into sediments is only about 50% -- well within the limits of our knowledge of that flux from the concentrations of organic matter in rocks.

The geologic behavior of sulfur, as revealed by the isotope ratios, has been given to illustrate the massive feed-back mechanisms that have operated through geologic time, and which promise (eventually!) to rectify most of man's current contributions to imbalances in the natural cycles. The major area of ignorance is in the rates of operation of the feed-backs. The rates are of critical importance, because if they are slow, compared to man's ability to interfere seriously with natural fluxes within a few years or a few decades, they may 'catch up' with his activities too late to be helpful. It is hoped that these rates will soon be investigated in detail.

## Sulfur Isotopes at Salt Lake City

The use of sulfur isotope ratios in long-term cycling studies has been discussed; now their current importance can be shown by a study of atmospheric sulfur gases in Salt Lake City, Utah.* Like most cities, Salt Lake has problems with high concentrations of sulfur gases. To complicate the situation, the city is a center of active mining of metals that occur as sulfide minerals, and smelters (as well as other kinds of industry and automobiles) are sources of sulfur gases. By studying the isotopic composition and the concentrations of sulfur gases, Jensen was able to separate man's contributions from those of nature.

Jensen found 3 major sources of sulfur in the Salt Lake City atmosphere: sulfur from oil, both from refineries and from automobile exhaust, with a $\delta^{34}S$ of about +16; sulfur from copper smelters, with $\delta^{34}S$ close to one; and bacteriogenic sulfur released from the Great Salt Lake as $H_2S$, with $\delta^{34}S$ averaging +5.3.

The mean value of all these sources, during 'normal' conditions, with refineries, automobiles, and smelters active, was +1.5 in the area close to the smelters, and +3.1 in the city proper. The initial implication of such low values was that the smelters were the chief contributors to atmospheric sulfur. If control were by automobile exhaust and refineries, plus general atmospheric contribution, $\delta^{34}S$ should be about +16. Bacteriogenic sulfur has a relatively low value (+5.3), but was not originally considered as a significant source.

During July, 1971, the smelter workers went on strike, so that there were no emissions for the several weeks that the smelter was closed. Values of $\delta^{34}S$ rose to +5.3 in the smelter area, and to +6.4 in the city proper. The expected values were perhaps +9 (general atmospheric contribution) in the smelter area, and as high as +16 in the city.

The conclusion was inescapable that the major source, while the smelters were not operating, was bacteriogenic sulfur from the Great Salt Lake, and that

---

* Jensen, M.L., 1972. Bacteriogenic sulfur in air pollution, Science 177, 1099-1100.

refineries, automobiles, and general atmospheric sulfur caused trivial additions. The value of $\delta^{34}S$ in the smelter area (within the limits of analyses and averaging) became the same as that of bacteriogenic sulfur, whereas that in the city rose only 1% above the bacteriogenic level.

From Jensen's data, approximate calculations can be made of the relative importance of sulfur from the smelters, from bacteriogenic sources, and from refineries and automobiles. The mixture of smelter sulfur + bacteriogenic sulfur, during smelter operation, has a $\delta^{34}S$ of +1.5. Using the values for the individual sources:

$\delta^{34}S$ (mass of smelter emission) + $\delta^{34}S$(mass of bacteriogenic emission) = $\delta^{34}S$ (mass of smelter emission + mass of bacteriogenic emission),

1.0(mass of smelter emission) + 5.3(mass of bacteriogenic emission) = +1.5(mass of smelter emission + mass of bacteriogenic emission).

Solving,

$$\frac{\text{mass bacteriogenic sulfur}}{\text{mass smelter sulfur}} = 0.13, \text{ or } 13\%.$$

During the strike, the $\delta^{34}S$ was +6.4. Assuming that this value is a mixture only of sulfur from refineries and automobiles and of bacteriogenic sulfur permits us to write:

+16(S from refineries and autos) + 5.3(bacteriogenic S) = +6.4(S from refineries and autos + bacteriogenic S).

Solving:

$$\frac{\text{S from refineries and autos}}{\text{S from bacteriogenic sources}} = 0.12, \text{ or } 12\%.$$

Thus, in the city, sulfur from the Great Salt Lake is about 10 times as important as that from man's additions.

Finally, the relative importance of emissions in the city, during smelter operation, can be calculated:

+16[$M_1$(mass of sulfur from refineries and autos)] + 5.3[$M_2$(mass of bacterial S)] + 1.0[$M_3$(mass of smelter S)] = 3.1[$M_1 + M_2 + M_3$].

From the relations above,

$$M_1 = 0.12 \, M_2.$$

Substituting:

$$+16(0.12M_2) + 5.3(M_2) + 1.0(M_3) = 3.1(0.12M_2 + M_2 + M_3).$$

Solving:
$$M_2 = 0.54M_3.$$

Salt Lake City, under usual conditions, receives twice as much sulfur from the smelters as from bacteriogenic sulfur coming from Great Salt Lake, and 20 times as much sulfur from the smelters as from refineries and automobiles.

## Summary

Jensen's study of Salt Lake City shows how powerful the use of isotopes can be in tracking down pollution sources. With both isotope and detailed emission data, the sulfur budget for a given area could be worked out in detail, giving a firm base for attack on environmental problems. Other isotopes, notably those of carbon, hydrogen and oxygen, have comparable use and potential.

Study Questions

1. Are any of the sulfur gases emitted to the atmosphere by man's activities likely to accumulate in the atmosphere?

2. So far as local pollution is concerned, are sulfur gases more dangerous than nitrogen gases?

3. Why is there currently great concern about sulfur gas emissions?

4. Can sulfur emissions from coal burning or from smelters be controlled?

5. What are some current problems in the use of catalytic emission control devices for $CO$ and $NO_X$ gases?

6. If $SO_2$ in the atmosphere oxidizes to $H_2SO_4$, then why is so much of the sulfate rained out to the earth's surface classified as $XSO_4$, instead of $H_2SO_4$?

7. In Bruges, Belgium, there has been rapid recent deterioration of the external surfaces of many buildings. What is the role of sulfur in this process?

8. Why the discrepancy between the source of $SO_4$ in streams between estimates by atmospheric scientists and geologists?

9. What is a possible explanation of this anomaly?

10. Cite a specific study that is evidence for the above interpretive answer.

11. Name two well-recognized locales of bacteriogenic additions of $H_2S$ to the atmosphere.

12. When pyrite in rocks is weathered, does all the sulfur oxidize to sulfate in the soil, and thus get transported into streams?

13. If, as is commonly assumed, all $SO_4$ in streams was derived from solution of gypsum and oxidation of pyrite, how could sulfur isotopes be used to find the relative proportions from each source?

14. Why can ancient gypsum deposits be used as a measure of the $\delta^{34}S$ of the contemporaneous ancient oceans?

15. If the long-term geologic sulfur system were truly steady state, would there have been any variation of the $\delta^{34}S$ of gypsum deposits as a function of their ages?

16. What does the geologic record of $\delta^{34}S$ as deduced from gypsum deposits show about the assumption of a complete steady state in the geologic past?

17. What other elements must have participated in this transfer?

18. Does the above create a problem for atmospheric oxygen during Permian time?

19. What might have been a feed-back mechanism to maintain atmospheric $O_2$ at survival levels?

20. What is the present major sink for pollutional atmospheric sulfur, and is it adequate to prevent atmospheric accumulation?

21. Where are two sites where the effects of man-made atmospheric S are actually felt?

# THE NITROGEN CYCLE

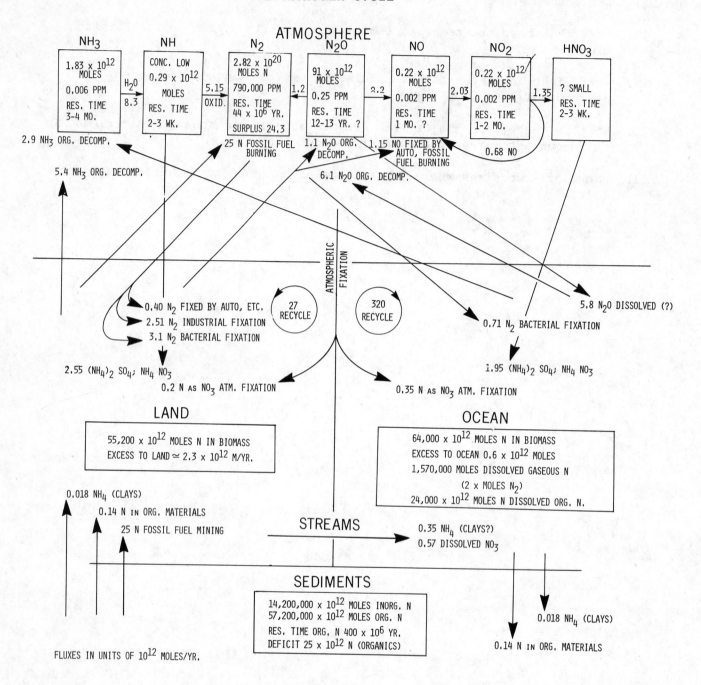

**FIGURE 32**

MANY SOURCES - SEE ESPECIALLY C.C. DELWICHE, 1970, THE NITROGEN CYCLE, SCIEN.
AMER., V. 223, P. 3-12; AND E. ROBINSON AND R.C. ROBBINS, 1970, GASEOUS ATMOS-
PHERIC POLLUTANTS FROM URBAN AND NATURAL SOURCES. IN GLOBAL EFFECTS OF ENVIRON-
MENTAL POLLUTION, S.F. SINGER, ED., SPRINGER VERLAG, NEW YORK, P. 55-59.
(THEIR FLUXES DIFFER FROM MOST OTHER ESTIMATES BY 10x).

Chapter 8


THE NITROGEN CYCLE


The nitrogen cycle is remarkably complicated and therefore many aspects of it are not well known.  There are 7 atmospheric species that must be considered in making balances.  On the land and in the sea the cycle of nitrogen-containing organic compounds is controlled by a complex system of bacterial oxidation and reduction.  The situation is further complicated by man's interference, both in terms of addition of nitrogen compounds to the atmosphere by fossil fuel burning, and by large-scale industrial fixation of atmospheric nitrogen.

## The Atmosphere

Figure 32 shows atmospheric nitrogen reservoirs and fluxes, and is used as a basis for the ensuing discussion.  It is convenient to start at the left side of the diagram with ammonia ($NH_3$).

Ammonia is soluble in water;  its atmospheric content averages about 0.006 ppm (6 ppb).  This concentration gives a total mass of $0.006 \times 10^{-6} \times 52 \times 10^{20}$ g (atmospheric mass) = $31 \times 10^{12}$ g, or $31 \times 10^{12}$ g/17g/mole = $1.83 \times 10^{12}$ moles. Ammonia is produced by the bacterial decomposition of organic materials.  The flux to the atmosphere is quite uncertain;  estimates here are of $5.4 \times 10^{12}$ moles/yr from the land, and $2.9 \times 10^{12}$ moles/yr from the sea.  These fluxes, in comparison to the size of the atmospheric reservoir, give a residence time of a few months.  In the atmosphere, ammonia is hydrolyzed by contact with water vapor to ammonium ions ($NH_4^+$) and hydroxyl ions ($OH^-$), resulting in the total flux of $8.3 \times 10^{12}$ moles/yr from the ammonia reservoir to the $NH_4^+$ reservoir.  The conversion of $NH_3$ to $NH_4^+$ is surprisingly slow, considering the high solubility of $NH_3$ in $H_2O$.  Once $NH_4^+$ is generated it is removed rapidly.  The reservoir is small (only $0.29 \times 10^{12}$ g), and the residence time is short (a week or several weeks).

Removal of $NH_4^+$ from the atmosphere is by rainout as ammonium sulfate or as ammonium nitrate [$(NH_4)_2SO_4$ or $NH_4(NO_3)$], or possibly by oxidation to $N_2$. Little is known about the relative rates of these processes.  Here it has been assumed that all nitrate generated in the atmosphere (far right box) combines with $NH_4^+$ to make ammonium nitrate, and that the $NH_4^+$ that combines with $SO_4^=$ is appreciable ($1.0 \times 10^{12}$ moles/yr). The numbers given here are subject to major change, but as will be seen, are fairly consistent with the requirements of an overall balance. Reference back to the sulfur cycle diagram shows that sulfate is rained out of the atmosphere as $H_2SO_4$ and as $XSO_4$. It is known that $NH_4^+$ is an important species in the $XSO_4$ category.

Next consider the third box from the left, atmospheric nitrogen. Molecular nitrogen, $N_2$, is unique among the species that have been considered so far, in that a major percentage of total N in the exogenic cycle is in the atmosphere. Almost all other elements have their large reservoirs in sedimentary rocks. To become available for cycling in the exogenic system, $N_2$ must be 'fixed', or converted to species with higher or lower valences. It was not until the late 1800's that industrial methods for fixing nitrogen were developed. Since then, however, because of the great demand for fixed nitrogen compounds as fertilizers, industrial production has increased enormously, until at present the number of moles fixed industrially rivals the amount fixed naturally by bacterial processes (estimated $2.51 \times 10^{12}$ moles/yr versus $3.1 \times 10^{12}$ moles/yr). Various atmospheric processes also fix nitrogen, for example lightning discharges can convert it from the molecular form (an estimated $0.55 \times 10^{12}$ moles/yr). Very broadly speaking, there is little worry about using up the $N_2$ in the atmosphere. Its residence time is estimated at 54 million years, so processes such as industrial fixation should not be significant, even at the highest estimates of increases in industrial fixation rates.

Nitrous oxide (laughing gas) is more or less a blank in our data. Its atmospheric concentration is high with respect to all other atmospheric species except $N_2$ so it has the second largest reservoir of atmospheric N gases. It is chiefly produced by decay (in addition to $NH_3$). Fluxes from land and ocean are poorly known. The estimate here of $7.2 \times 10^{12}$ moles/yr gives a residence time of 12-13 years.

$N_2O$ is converted to NO and $N_2$ by photochemical reactions in the stratosphere. The concentration of NO is only about 0.002-0.003 ppm, and the residence time on the order of one month. NO is oxidized to $NO_2$ chiefly by ozone, resulting in chemical destruction of ozone:

$$NO + O_3 = NO_2 + O_2$$

$$O_3 + hv = O_2 + O$$

$$NO_2 + O = NO + O_2$$

$$\text{net: } 2O_3 + hv = 3O_2,$$

where hv is a unit of solar radiation.

NO acts as a catalyst; nitrogen oxides are present in the stratosphere in amounts sufficient to destroy 70% of the naturally produced ozone each year!

As shown in Figure 32, production of nitrogen oxides by man's activities is initially attributed entirely to NO, with oxidation to $NO_2$ occurring in the atmosphere. At present, stationary-source fuel combustion accounts for about half of nitrogen oxide emissions and electric power generating plants produce half of this amount. About 35% of the flux comes from transportation sources; the temperatures of gasoline-burning piston engines are high enough to produce nitrogen oxides by fixing atmospheric nitrogen. It is these oxides, plus other combustion products and solar radiation, that give the complex set of nitrogen oxides that are so toxic and irritating in smog.

Both subsonic and supersonic aircraft (SST's) can produce nitrogen oxides and consequently lead to destruction of stratospheric ozone (100 supersonic jetliners could reduce ozone by between 0.23 and 2.1%). For each 1% reduction in ozone, about 2% more ultraviolet radiation could reach the earth's surface. Recently, chlorinated hydrocarbons have been measured in the troposphere; their 'escape' to the stratosphere could also lead to destruction of ozone by catalytic reactions similar to those for nitrogen oxides.

$NO_2$ is one of the few colored gases. Most people have seen the reaction of nitric acid with a copper penny to produce this brown gas. $NO_2$ reacts with water to give nitric acid and NO:

$$3NO_2 + H_2O \rightarrow 2HNO_3 + NO.$$

This reaction explains the feed-back loop on the upper right of Figure 32; $NO_2$ is 'disproportionated' to produce $HNO_3$ and NO, so the NO returns to the NO reservoir and progresses again through $NO_2$ to $HNO_3$.

The fate of $HNO_3$ is not clear. Certainly some of the $HNO_3$ formed reaches land and sea as $HNO_3$. In the balance presented, all of it is reacted with $NH_4^+$

or with other cations.  Only time and more chemical analyses will reveal the actual fluxes.

## Toxicity

Ammonia is essentially non-toxic, nor are ammonium compounds a problem. Molecular $N_2$ is, for people, only a diluent of other gases they may need (like $O_2$) or not need (like $SO_2$).  Nitrous oxide, commonly used as an anaesthetic, can be tolerated in very high concentrations with no toxic effects.  Its effects in the human body are dissipated within a few hours.

NO is not a problem because of its very low concentrations;  $NO_2$ is a gas of a different color and effect.  It is highly toxic and irritating;  the current maximum permissible level is about 5 ppm.

The currently accepted upper limit for NO is 25 ppm, so its short residence time and low concentration as compared to $NO_2$ relegates its importance in pollutional problems to its role as a precursor of $NO_2$, and not to its direct toxic effects.

## Land, Sea and Rocks

Figure 32 also gives current estimates of nitrogen reservoirs and fluxes among land surface, ocean, and sedimentary rocks.  There are great uncertainties in dividing fluxes from the atmosphere between land and sea.  Actually, this has been done in the figure, but little credence should be given to the estimates.

Starting with the land in Figure 32, it should be noted that the number of fluxes to and from the surface is large.  The natural fluxes caused by the continual uplift and erosion of the land are relatively small.  It is known that $NH_4^+$ is absorbed by clays, but the estimate given of that flux is highly suspect.  The N content of organic materials is a better number and is derived from the average content and composition of disseminated organic materials in rocks.  The flux of molecular nitrogen to the atmosphere reflects only the $N_2$ added to the atmosphere by fossil fuel burning.

In terms of the fluxes from the land surface to the atmosphere, the estimates are based on the same kinds of evidence as given for the reverse fluxes from the atmosphere.  However, a crude estimate has been made that the ammonium sulfate and ammonium nitrate received by the land are slightly larger than that received by the sea, despite the difference in their surface areas.  This estimate considers that most nitrate is probably pollutionally derived, and consequently is preferentially returned to the land because of its short residence time.

Both on land and in the sea, the photosynthetic production of organic material is large. As shown before, it represents a total of the order of 5000 x $10^{12}$ moles of carbon/yr, or, from the average composition of organic material, about 350 x $10^{12}$ moles/yr N fixed by photosynthesis. This takes place on land and in the sea, by a cycling of N; fixation by photosynthesis, release by decay. As compared to carbon, for which exchange during photosynthesis is between atmosphere and land, N cycling takes place within the soil or in ocean water. The fixation and release within these cycles are much greater than the fluxes from land or sea to the atmosphere. The basic chemical reactions involved are given below, as well as the schematic cycling (Fig. 33) of N between organisms and inorganic species.

Bacterially controlled nitrogen reactions<br>and the N cycle in soil and water

(Type of bacterium responsible for reaction is indicated)

nitrogen-fixing bacteria<br>(Azobacter)

$$2N_2 + 6H_2O = 4NH_3(\text{organism}) + 3O_2$$

decay

$$4NH_3(\text{organisms}) = 4NH_3 \text{ gas} + \text{other decay products}$$

hydrolysis

$$4NH_3 \text{ gas} + 4H_2O = 4NH_4^+ + 4OH^-$$

bacterial nitrification<br>(Nitrosomonas)

$$4NH_4^+ + 6O_2 = 4NO_2^- + 8H^+ + 4H_2O$$

bacterial nitrification<br>(Nitrobacter)

$$4NO_2^- + 2O_2 = 4NO_3^-$$

bacterial denitrification<br>(Pseudomonas)

$$4NO_3^- + 2H_2O = 2N_2 + 5O_2 + 4OH^-$$

assimilation

$$4NO_3^- + 8H_2O = 4NH_3 \text{(organisms)} + 8O_2 + 4OH^-.$$

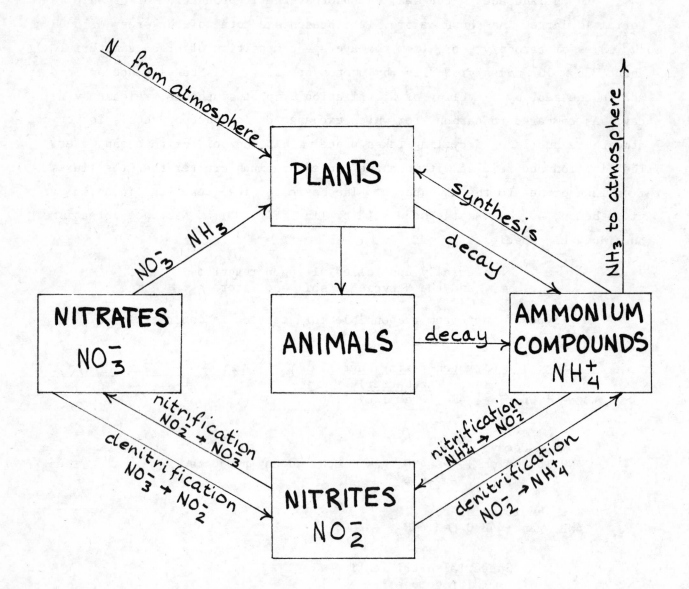

SCHEMATIC N CYCLE
Figure 33

The land balance in Figure 32 shows an excess of about $2.3 \times 10^{12}$ moles/yr N delivered to the land. This seems to reflect a real situation, in which the addition of nitrate in fertilizers and from automobiles has not entirely shown up in increased $NO_3^-$ in streams. Note that the stream-transported $NO_3^-$ is apparently still very small, despite recent large increases in the amount of N applied as fertilizer. Presumably some of this nitrogen is still making its way thorugh soils to groundwater, and eventually will appear in streams.

This excess of $2.3 \times 10^{12}$ moles/yr N may conceivably be related to an increase in the biomass, which was suggested before as a solution to the problem of the $CO_2$ that has been emitted to the atmosphere from fossil fuel burning, but which is no longer there. In fact, the annual accumulation of N on the land would account for about 66% of the $CO_2$ that seems to have disappeared from the atmosphere if both N and $CO_2$ were fixed in the biomass.

Delivery of N to the oceans via streams is small, compared to the atmospheric fluxes in and out of the ocean. The arbitrary division of $NH_4NO_3$ and $(NH_4)_2SO_4$ fluxes between land and ocean involves a total flux to land three times greater than the stream flux, so that the balance for moles of N in and out of the oceans is highly suspect. However, Figure 32 shows an annual surplus of $24.3 \times 10^{12}$ moles of $N_2$ in the atmosphere, and an excess of $2.3 \times 10^{12}$ moles/yr on land, plus that in the sea. Just how this rough balance is related to man's additions of $N_2$ and fixed nitrogen to the system (0.4 from automobiles, and 2.51 by industrial processes drawing on the atmosphere) is not clear.

In summary, the best that can be said is that despite the large amounts of nitrogen in the various reservoirs, the critical function of nitrogen as a nutrient depends upon relatively small fluxes based on bacterial activity. Man's fixation of nitrogen by automobiles and by industry has developed fluxes comparable to those accomplished in nature by bacteria. The current situation is one that needs much more investigation, because the degree of interference with the natural system is a high percentage of natural fluxes, and the whole photosynthetic system, and hence the survival of life, depends upon these fluxes. The nitrogen reactions may be the most important of all considered here, in terms of deciding whether man's activities will enhance his future well-being or have jeopardized it seriously.

### Oxygen, Phosphorus and Nitrogen

In Chapter 5 it was shown that there are feed-back mechanisms between photosynthetic rate, burial of organic compounds, and oxygen content of the atmosphere. In that discussion, organic matter was represented simplistically as $CH_2O$. In fact, to make plant and animal materials, many elements are required, among them nitrogen, phosphorus, and sulfur.

$CO_2$ for photosynthesis comes from the atmosphere, but how about the other required elements? Nitrogen is abundant in the atmosphere, but it exists as molecular nitrogen, and is not directly available for the photosynthetic process. It must be converted into other forms before it can be used, and the conversion processes are bacterially controlled, so that they take place in the soils or in the sea. If there is not enough nitrogen in the proper form, the amount of organic material that can be photosynthesized is limited by this nutrient nitrogen.

The same nutrient demand exists with respect to phosphorus. Phosphorus compounds are non-volatile, with the possible exception of phosphine ($PH_3$), a reduced form of the element. But phosphine is rapidly oxidized to phosphates ($PO_4^{-3}$), which are available to plants only in dissolved form in soils or in the sea. Thus available phosphate can also be a limiting nutrient, and its availability is independent of any atmospheric content. The argument could be raised that photosynthetic rates, and hence the amount of new organics deposited, are in fact controlled by nitrogen and phosphorus availability, and that it cannot be assumed that the rate would be constant, which would affect the burial of new organics and atmospheric oxygen. It is generally accepted that photosynthetic rate is indeed limited by phosphorus and nitrogen availability. Current hypotheses say that in the ocean nitrogen is the limiting nutrient; on land it is phosphorus. Thus the constancy of the percent organic content of sedimentary rocks through time indicates that the rate of nutrient addition and subtraction from the oceans has been, on the average, nearly constant. Furthermore, oxygen content of the atmosphere could well be changed by man's significant addition of the limiting nutrients, P and N. However, models taking into account the present and projected rates of addition of N and P to the exogenic cycle lead to the preliminary conclusion that changes of atmospheric oxygen of more than 1% will require hundreds of thousands of years. The effect of adding nutrients to the cycle would cause an increase in $O_2$.

Study Questions

1. The ratio of C/N in terrestrial living organisms is about 100/1 (as opposed to a ratio in soil humus or in buried marine organic matter of 12-15/1). What is the number of moles of N fixed in living terrestrial plants each year?

2. Figure 32 (N cycle) shows only $3.1 \times 10^{12}$ moles N/yr fixed on land by bacteria (and affiliated leguminous plants). How is it possible, then, that $25 \times 10^{12}$ moles of N/yr are taken up by terrestrial plants?

3. From the above, what is the approximate % loss of N from the natural soil cycle per year?

4. Is the change in the mole ratio of C/N from 100/1 in living plants to about 14/1 in soil humus consistent with the answer to question #3?

5. Nitrogen fixed commercially, used for fertilizer, is roughly equivalent to nitrogen fixed naturally. Does this mean that the terrestrial biomass should have doubled?

6. From Figure 32, what is the total 'fixed' N added to the oceans/yr?

7. If the mole ratio of C/N in the living biomass of the oceans is 6/1, what is the annual loss of N from the oceanic surface system?

8. How does this loss compare to N fluxes from the ocean?

9. If fertilizer N goes chiefly into crops consumed by animals and then via animals to man, what is its eventual fate?

10. Why should there be little concern about buildups of NO and $NO_2$ (the chief gases from automobiles) in the global troposphere?

11. The $N_2$ reservoir (Fig. 32) shows a surplus of $24.3 \times 10^{12}$ moles $N_2$/yr. If this surplus is real, should there be concern about the buildup of $N_2$?

12. Information is poor on $N_2O$ fluxes. Is it possible that the residence time listed (12-13 years) could actually be only 3 weeks, like most of the other gases?

13. Nitrogen is clearly the limiting nutrient in the Pacific Ocean, but both N and P are limiting in the N Atlantic. Is there a possible explanation for these relations, based on man's activities?

14. Which of the N gases are worrisome as local pollutants?

15. Is there concern about the increase of $NO_3^-$ in groundwater and streams in terms of its toxicity?

16. Outline the chemical reactions that result in much fixed N coming to the land and sea as $(NH_4)_2SO_4$ and $NH_4NO_3$.

17. If all the N fixed as fertilizer ($2.51 \times 10^{12}$ moles/yr) were used in lakes to form water plants with a ratio of C/N of 100/1, how many moles of organic material would be formed?

18. At 30 g/mole, how many grams of organic material does this (Question 17) represent?

19. The area of the Great Lakes is about $2600 \times 10^{12}$ $cm^2$. If the density of organic material is 1.0, how thick a layer of organics would have to accumulate each year to remove all the fertilizer N?

# THE PHOSPHORUS CYCLE

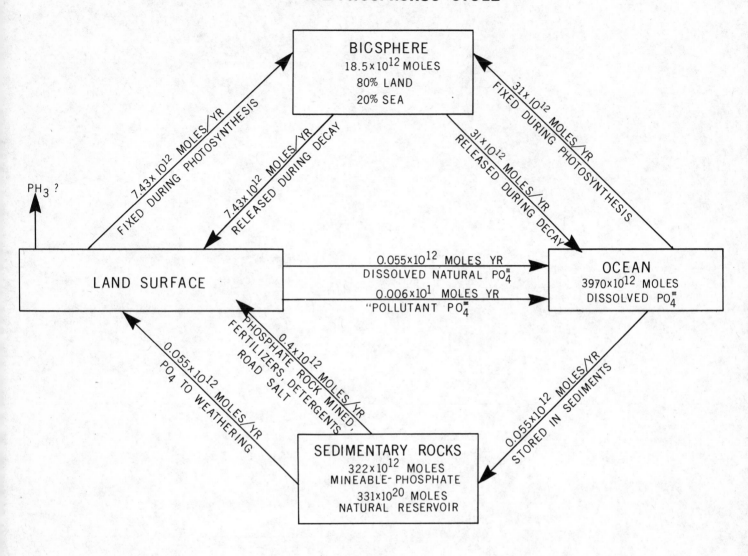

**FIGURE 34**

Chapter 9

THE PHOSPHORUS CYCLE

A simplified phosphorus cycle is discussed here; for a recent and more quantitative interpretation see Chapter 12. Comparison between the two cycles shows how reservoir masses and fluxes change as data become available, and provides a feeling for the difficulties faced in making such estimates.

The phosphorus 'cycle' is shown in Figure 34. Phosphorus compounds are not only non-volatile, so that they play no significant role in the atmosphere, but are also relatively insoluble in water. The major reservoir of phosphorus is in sedimentary rocks, where it occurs chiefly as calcium phosphate $[Ca_3(PO_4)_2]$. The low solubility of phosphorus is shown by its small flux as dissolved material in streams, only $0.055 \times 10^{12}$ moles/yr, as compared for example, to the flux of dissolved calcium, which is hundreds of times greater. The average composition of plant material is $C_{280}H_{560}O_{280}N_{19}P_1$; only one phosphorus atom per 280 carbon atoms. But the P atom is required, so phosphorus may still be a limiting nutrient because of its low solubility.

The publicity that phosphorus has received is related to the addition of soluble phosphorus compounds to streams and lakes, largely in the form of detergents and in sewage. As a result of such additions, in water bodies where phosphorus originally controlled photosynthesis, the water has been 'enriched', and organics (weeds, algae, water lilies, etc.) have proliferated. The lakes and streams become choked with plants, the plants die and sink, using up oxygen in the deeper waters, and the water bodies become eutrophic.

The question of whether P or N is the limiting nutrient is illustrated by the following sketches:

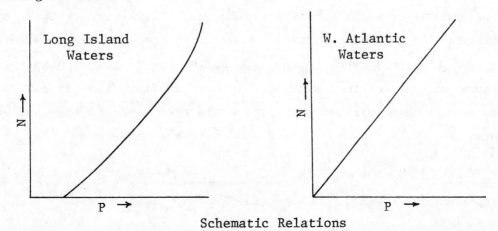

Schematic Relations

In Long Island water N becomes essentially zero while there is still significant dissolved phosphate, so N is the limiting nutrient. In the Western Atlantic, N and P approach zero linearly, so both are limiting. Where N is the limiting nutrient, growth of nitrogen-fixing organisms, such as the blue-green algae, often is prolific. Consequently, algal growth tends to be encouraged by high phosphate levels, as in water bodies into which phosphate detergents are discharged. In some instances, especially in fresh water lakes, if both P and N are high, $CO_2$ can become limiting; it becomes depleted below required photosynthetic levels (approximately .0001 atm) by plant growth, because growth exceeds the rate of addition of new $CO_2$ from the atmosphere.

A current issue is whether or not to discontinue the use of phosphate detergents. One school of thought would rely on preventing excess phosphate from entering water bodies at any stage; the other would rely on removal of phosphate at water treatment plants. One ingenious procedure is to use algae to remove the excess phosphate, and then to recover the phosphate from the algae. The method works well; the use of it depends on whether it can be economically applied.

Global effects of man's addition of P to the surface environment are difficult to estimate. Relatively few analyses have been made of the dissolved P in unpolluted streams. The natural level is very low ($\cong$ 0.01 ppm), even though it is still above the level required for use of P as a nutrient ($\cong$ 0.001 ppm). As shown in the diagram of the phosphorus cycle, current estimates are that increase in global stream phosphate as a result of man's activities is only a few percent. Like so many of the other pollutants studied, the effects are highly local and strongly correlated with urbanization.

The major addition of P to the natural cycle by man is in his use of fertilizers, but even their extensive use has apparently not affected the dissolved P of streams significantly. In fact, the fertilization process consists of adding large amounts of soluble phosphates to the soil. They are taken up in part by plants, but tend to become fixed as insoluble compounds -- aluminum phosphate ($AlPO_4$), and iron phosphate, $Fe_3(PO_4)_2$. During the cycles of plant growth and decay, more and more of the released phosphate is fixed inorganically. Eventually the phosphate is carried down streams in the suspended load and buried in sediments.

The rate of phosphate fertilizer addition to the land surface is much higher than the natural rate at which phosphates are exposed by uplift and erosion. Thus the minimg of phosphate deposits for fertilizer is comparable to the burning of fossil fuels -- the mined phosphate is 'consumed' by conversion to insoluble, non-recoverable compounds. Estimates vary, but phosphate for fertilizer may be one of the critical long-term resource problems, because world demands for higher food supply will require more and more fertilizer as soils become depleted. Perhaps the answer, in part, could come from methods of releasing inorganically fixed phosphate, but this is at present only a hope.

In the complete exogenic cycle, it appears that the iron and aluminum phosphates are converted to calcium phosphates by geologically slow processes.

Phosphate level in the oceans (below the organism-depleted surface water) apparently is controlled by the solubility of $Ca_3(PO_4)_2$, so that as we look at the long-term picture, productivity, in terms of phosphate limitation, depends upon the rate of supply of deep water phosphate to the surface, which is in turn a function of circulation and of the solubility of $Ca_3(PO_4)_2$; i.e., with the same circulation, higher solubility of $Ca_3(PO_4)_2$ would supply the surface waters at a higher rate.

Consequently, there may be a link between atmospheric $CO_2$ and phosphate availability for oceanic systems. If $CO_2$ rises, oceanic acidity is increased. This in turn would increase $Ca_3(PO_4)_2$ solubility:

$$Ca_3(PO_4)_2 + 4CO_2 + 4H_2O = 3Ca^{++} + 2H_2PO_4^- + 4HCO_3^-$$

One of the major problems that has been raised by studies of the carbon, nitrogen and phosphorus cycles is that of the sink for the $CO_2$ added to the atmosphere by fossil fuel burning that disappears from the atmosphere. The current increase in atmospheric $CO_2$ has been roughly constant for the last several years, at about 0.9 ppm, which represents an accumulation of about $107 \times 10^{12}$ moles/yr. The total $CO_2$ added by fossil fuel burning is of the order of $388 \times 10^{12}$ moles/yr. Thus some $280 \times 10^{12}$ moles of $CO_2$ are removed by some mechanism. It may go into the ocean, or it may go into increasing the mass of terrestrial living material. If we use our estimate of the composition of living terrestrial plants, we find that $280 \times 10^{12}$ moles of $CO_2$ would photosynthesize to living material requiring $3.15 \times 10^{12}$ moles of fixed nitrogen and $0.35 \times 10^{12}$ moles of phosphorus. It may be a coincidence, but these values are remarkably close to the nitrogen fixed industrially each year (about $2.5 \times 10^{12}$ moles/yr) and the phosphate mined for fertilizers and detergents (about $0.40 \times 10^{12}$ moles/yr), suggesting that at least some of the excess C, N and P added by man goes to increase the terrestrial biosphere. The conclusion that the excess $CO_2$ goes into living material is supported indirectly by the small amounts of N and P carried from the land to the sea by streams. These are estimated at $0.06 \times 10^{12}$ moles of P and $0.57 \times 10^{12}$ moles N ($NO_3^-$), small fractions of the N and P added to the land by man. Thus there would seem to be an accumulation of living matter on land in response to the addition of fertilizers that is compensating for a significant fraction of the $CO_2$ that we are adding to the atmosphere by burning fossil fuels. It may emerge, ironically, that man is inadvertently, despite his major interference in the C, N and P cycles by his technologic activities, coming close to indulging in activities that balance out his interference in these cycles.

Study Questions

1. What is one major difference between the phosphorus cycle and the carbon and nitrogen cycles?

2. What is the chief phosphorus mineral in sedimentary rocks?

3. What concentrations of F in natural waters are required for extensive sub-stitution of F for OH, and what public health controversy is related to the required concentration?

4. Has the extensive addition of P to soils in the form of soluble phosphates, and the use of soluble phosphates in detergents, increased the dissolved phosphate reaching the oceans significantly?

5. How can the above be true, if man's additions of P to the terrestrial surface environment are many times the natural fluxes?

6. Despite the apparently small interferences of man in the global P cycle, have his activities created important environmental problems?

7. What are some of the preventive measures that have been applied to these problems?

8. Can blue-green algae flourish even if only excess P is added to a lake, without increase in N?

9. What seems to be the basic control of the P in the oceanic reservoir?

10. What becomes of the iron and aluminum phosphates formed by addition of fertilizer to soils?

11. Although the specific effects are difficult to assess, would a marked change in climate have a greater effect on the oceanic biomass or on the terrestrial biomass, in terms of P availability?

12. Why has such a wide range of values been estimated for the mineable P reservoir?

Table 16

Comparison of mining production of some metals, metal emission rates to
atmosphere owing to man's activities, worldwide atmospheric rainout, and total
stream load.  (units of $10^{12}$g/yr)

| Metal | Mining | Emission | Atmospheric Rainout | Stream Load | Interference Index (Atmosphere) (Emission/Rainout) x 100 |
|---|---|---|---|---|---|
| Pb | 3 | 0.40 | 0.31 | 0.42 | 129% |
| Cu | 6 | 0.21 | 0/19 | 0.82 | 111% |
| V | 0.02 | 0.09 | 0.02 | 2.4 | 450% |
| Ni | 0.48 | 0.05 | 0.12 | 1.2 | 42% |
| Cr | 2 | 0.05 | 0.07 | 1.7 | 71% |
| Sn | 0.2 | 0.04 | -- | 0.27 | -- |
| Cd | 0.014 | 0.004 | -- | 0.04 | -- |
| As | 0.06 | 0.05 | 0.19 | 0.3 | 26% |
| Hg | 0.009 | 0.01 | 0.001 | 0.005 | 1000% |
| Zn | 5 | 0.73 | 1.04 | 1.8 | 70% |
| Se | 0.002 | 0.009 | 0.03 | 0.02 | 30% |
| Ag | 0.01 | 0.003 | -- | 0.03 | |
| Sb | 0.07 | 0.03 | 0.03 | 0.09 | 100% |

(see F.T. Mackenzie and R. Wollast, Sedimentary cycling models of global
processes;  in press in Vol. 6, The Sea, E.D. Goldberg, Ed., Wiley-Interscience, New York.)

Chapter 10

TRACE ELEMENTS

## Some General Comments

Trace elements, that is, those elements that usually occur in low concentrations in rocks, soils, waters, atmosphere and biota, have received considerable attention during the past decade because some are known to be toxic to the human body at low concentrations. Examples of such elements are mercury, lead, arsenic, cadmium, selenium and vanadium.

Because of lack of numerical data, few global models of trace element cycles have been developed; therefore, their environmental impact on a global basis is poorly known. However, because of man's activities, local concentrations of some of these elements have been multiplied many times, particularly in water. Illness and even death of many people have resulted.

On a global basis, an estimate of man's contribution of metals to their natural exogenic cycles can be made by comparison of mining production, emission rates to atmosphere owing to man's activities, world-wide atmospheric rainout, and total river load of the element (Table 16).

Calculation of the values for the rainout of metals from the atmosphere to the earth's surface provides some feeling for how the values in Table 16 were obtained. The calculation is based on estimates of the concentrations of metals in atmospheric particulates recovered from 'clean' air. These concentration values, reported in units of nanograms ($10^{-9}$g) per cubic meter (ng/m$^3$)* were then multiplied by the atmospheric volume up to an elevation of 5,000m. The standard assumption was then made that this volume of air is completely washed free of particulates 40 times each year; this is equivalent to a rain about every 10 days. For nickel, for example, the calculation is:

(concentration in air, 1.2ng/m$^3$)(volume of atmosphere to 5,000m,
510 x $10^{16}$cm$^2$ x 5 x $10^5$cm = 2.55 x $10^{22}$cm$^3$ = 2.55 x $10^{18}$m$^3$)
(rainouts/yr, 40) = 0.12 x $10^{12}$g Ni/yr.

---

* 1 nanogram/cubic meter $\cong$ 0.00077 ppb

Table 16 shows that mining production for many metals approaches the total stream load; indeed for the metals Pb, Cu, Cr, Hg and Zn, mining production apparently equals or exceeds the rate at which the metal is transported to the oceans in the dissolved and particulate load of streams. The emission rate of Pb and Hg to the atmosphere from man's activities already equals or exceeds the natural stream flux, and for the metals Cu, Sn, Cd, As, Zn, Se, Ag and Sb, emission rates are within an order of magnitude of the stream fluxes -- values well within the calculation errors (owing to poor data for emission rates and for concentrations of metals in the dissolved and particulate loads of streams) inherent in Table 16.

Comparison of emission rates to the atmosphere with atmospheric rainout is particularly informative. Regardless of the errors inherent in these calculations, it is evident from the interference indices in Table 16 that emission rates for all the metals shown are within an order of magnitude of rates of particulate fallout over the earth's surface, suggesting that metal input to the atmosphere via industrial activities, combustion of fossil fuels, etc., rivals natural inputs.

It is also interesting to note that some of those elements, such as Hg, As, Se and Pb that are considered detrimental to human health, are selectively emitted into the earth's atmosphere by natural (and anthropogenic) processes. Sedimentary rocks containing these elements are a major source of atmospheric particulates. Figure 35 illustrates that those trace elements with relatively low oxide boiling points (a measure of volatility) are more concentrated in the atmosphere with respect to their content in sedimentary rocks than are elements with high boiling points. In other words, elements like Se, As and Cd have high volatilities; elements like Pb, Zn and Ni have moderate volatilities; and Al, Ti and Sc are present in the atmosphere at concentrations similar to those in their crustal source. Figure 35 also shows that significant concentrations of some high volatility elements are present in atmospheric particulates at lower concentrations than would be expected if oxide boiling points are a reasonable index of potential volatility. These elements are probably present in the atmosphere in a state other than particulate -- possibly gaseous -- and have not been measured. Recently, mercury in the atmosphere over Tampa Bay, Florida, was observed to be principally in 'volatile' form, present as gaseous mercuric compounds, methylmercuric compounds, and metallic mercury

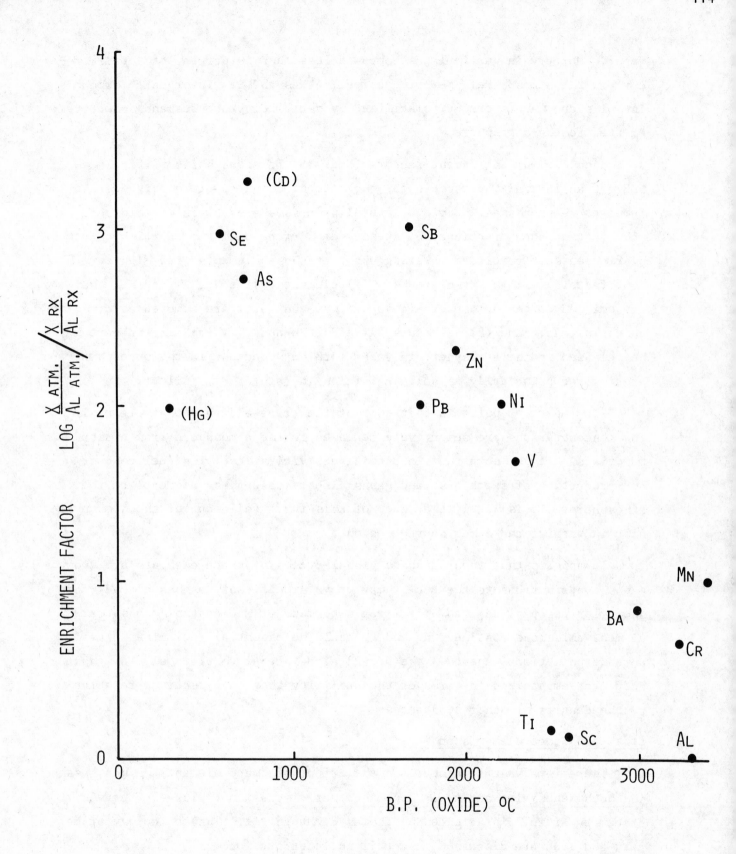

FIGURE 35.   VOLATILITY OF TRACE ELEMENTS TO THE ATMOSPHERE.

B.P. = BOILING POINT

vapor.  Mercury in particulates composed <u>less</u> than 10 percent of total mercury in the atmosphere.  Mercury, as well as As and Se, in organic compounds
in soils and waters can be volatilized by bacterial processes and selectively
emitted to the atmosphere.

High element enrichment factors (Fig. 35) in atmospheric particulates
suggest an initial vapor phase for the element owing to either high- or low-
temperature processes.  Obvious natural processes are volcanism, biological
mobilization, and fractionation at the sea surface during production of atmospheric sea salt particles.  We suggest that those elements with high enrichment factors fall into two groups:  (1) elements, like Hg, As and Se, that
are emitted to the atmosphere in vapor form and later are removed as dissolved
gases in rain;  and (2)  elements, like Pb, Zn and V, which also in part may
be released to the atmosphere in vapor form but condense in the atmosphere and
are removed principally as solid particles in rain or dry fallout.

The important point is that some of the trace elements that are toxic at
low concentrations are selectively released to the atmosphere from crustal
materials.  It is known that industrial activities involving fuel combustion
and production of stack dust and gases also release these elements to the
atmosphere.  It is likely that much of this input falls out of the atmosphere
near industrial and other source regions.

It appears that when adequate global cycles of trace elements are known,
man's contributions to the cycles may prove significant, as has already been
shown for mercury and lead.  If efforts to monitor and control man's trace
element emissions continue, Hg and Pb could be essentially eliminated as important pollutional agents.  The perils associated with trace elements have
been over-emphasized in terms of the publicity they have received relative
to the dangers of other substances.

Effect of Trace Elements on Man

There is a vast literature on the effect of trace elements on man, assessing both beneficial and toxic aspects.  Our discussion is largely derived from
a summary article in <u>Science</u>* by Thomas H. Maugh, II.  The toxicology of mercury and lead are discussed in detail in later sections.

_____

*Maugh, T.H., II, 1973.  Trace Elements:  A Growing Appreciation of Their
Effects on Man.  Science <u>181</u>, 253-254.

Trace elements can be beneficial or harmful to human health. Iron and zinc are essential to human metabolism; lead and cadmium are toxic at the same levels at which iron and zinc are beneficial; selenium is beneficial or toxic within a narrow range. The great complexity of interactions between trace elements in the environment and human health has been recognized only recently. Maugh states, "The time may thus be fast approaching when evaluation of trace element concentrations will play a fundamental role in the diagnosis of illness and when manipulation of these concentrations may play an even greater role in its prevention."

Correlation of disease with trace element concentrations caused the U.S. Public Health Service to place concentration limits on U.S. public water supplies. Permissible concentrations of elements in water supplies, plotted against average concentration in the world's rivers, are shown in Figure 36. Elements toxic at low concentrations are found at low concentrations in the world's rivers, and vice versa, which has interesting implications with respect to the adaptation of organisms to their environment.

The detrimental effects of some trace elements and their compounds on human health are well known. Lead and methylmercury damage the central nervous system: the severity of the damage is well documented in the episodes of methylmercury poisoning in Minimata Bay, Japan, and more recently in Iraq. Beryllium is carcinogenic and poisonous; cadmium, arsenic, selenium and yttrium have been shown to be carcinogenic to laboratory animals. Lead, copper, zinc and chromium may increase dental caries. It is established that incorporation of selenium during tooth development inhibits tooth mineralization and increases susceptibility to decay. Molybdenum and cadmium apparently affect bone development; molybdenum has been shown to reduce calcium uptake in rats, leading to osteoporosis, weakening of bone structure. The syndrome itai-itai ('ouch-ouch'), a severe and painful decalcification of bone, accompanied by multiple fractures, has been attributed in the Kinzu River area of Japan to cadmium discharge from a factory to the river, and ingestion of cadmium by the local residents.

The toxicity of certain trace elements may not be related to their specific toxicity but may be due to their interference with the function of other elements. Molybdenum inhibits the absorption of copper from food, producing copper defi-

117

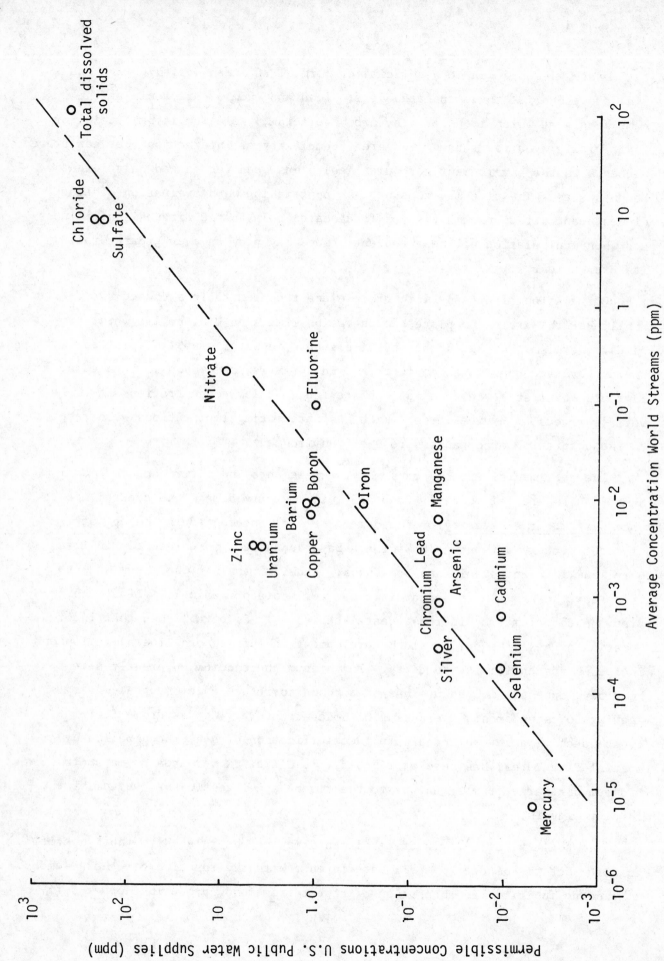

Figure 36: Correlation between trace element concentrations in streams and permissible concentrations in water supplies.

ciency. Addition of 10 ppm selenium to a low protein diet hindered growth of rats, produced enlargement of heart and liver, and lowered the concentrations of magnesium, copper and manganese in these and other organs. Effects were magnified when selenium was augmented with cobalt. R.E. Burch of the Veterans Administration (in Maugh, op. cit.) claims that the symptoms produced in rats by the combined addition of selenium and cobalt are similar to those observed in the 'beer drinker's cardiomy-apathy' that caused deaths in Omaha, Minneapolis, and two foreign cities in 1965 and 1966. The syndrome had previously been attributed to cobalt added to beer to stabilize its foam, but Burch concluded that the added cobalt may have induced toxicity from naturally-occurring selenium in the brewing water. This case is an excellent example of what can happen when interacting substances are added to food and drink.

Fourteen trace elements have been identified as essential to human metabolism. Some metabolic enzymes (chemical catalysts) function only if certain trace elements (such as cobalt, zinc or manganese) are present. Iron, of course, is an integral component of hemoglobin. Tooth enamel contains at least 43 elements. Some of these probably play no role in the prevention or development of caries; but fluorine, molybdenum, and perhaps vanadium, boron, strontium, barium, lithium, titanium and aluminum may help to prevent decay by stabilization of tooth structure. Claims have been made that addition of selenium and vitamin E to breakfast cereals accounts for the 70% decline in gastric cancer in the United States and Canada since the 1930's. Also, it appears that the incidence of cardiovascular disease is lower in areas where municipal water supplies contain high concentrations of dissolved minerals.

R.H. Seaply (in Maugh, op. cit.) of the Kettering Medical Center contends that despite the growing body of knowledge about trace elements and their association with illness, there has been little application of research findings to clinical medicine. There is growing evidence that routine analysis of trace element concentrations in blood may be useful in diagnoses, but is a little-used medical technique.

It has been found that stress from operations, injuries and burns increases concentrations of zinc in blood. The elevation is maximum during stress and abates when stress declines. Thus analysis of zinc concentrations could be used to monitor recovery from critical injury. Also, monitoring certain trace metals in the blood may aid in diagnosis of some types of heart disease. For example, manganese and nickel concentrations in blood rise sharply just before the onset of myocardial infarctions. It is possible that monitoring of these concentration changes could help in early diagnosis.

There remains much to be learned, particularly about interferences and interactions among trace metals and their combined effects on human physiology. Equally clear, however, as Maugh so succinctly states, "...is the realization that much of the research on trace metals may be to little avail if there is not a narrowing of the gap between research results and clinical applications and an increased effort to place the new knowledge in the hands of practicing physicians".

## Mercury, Lead and Manganese as Examples of Trace Element Behavior

Mercury, lead and manganese are examined here in some detail as representative trace elements. Their occurrence and general behavior are basically related to their relative volatilities (Fig.35 ) and are like those of a number of other trace elements -- arsenic, cadmium, selenium, vanadium and antimony.

### Mercury

Estimates of the pre-man and present cycles of mercury are given in Figures 37 and 38. Details of the calculations involved are presented in Chapter 12. It should be emphasized that the fluxes shown have large uncertainties because of difficulties in obtaining accurate values from the very low concentrations and wide ranges of concentrations of mercury in natural materials. Values for mercury in various substances are given in Table 17.

Mercury mines occur in relatively few places; foremost among them are the Almaden mines of Spain. In these deposits mercury occurs chiefly as the mineral cinnabar, HgS, and as native (metallic) mercury.

Our estimate of the pre-man flux of mercury to the oceans from weathering of rocks is:

$$130 \times 10^{-9} \times 100 \times 10^{14}g = 13 \times 10^8 g/yr,$$

using 130 ppb as the average mercury content of rocks and $100 \times 10^{14}$ g/yr as the total mass of rock weathered per year before man. Figure 37 assumes that the pre-man transport rate of mercury to the oceans equalled its depositional rate, which in turn was equivalent to the rate of mercury transport from uplift and erosion.

Major transfer in the pre-man sedimentary cycle was between the atmosphere

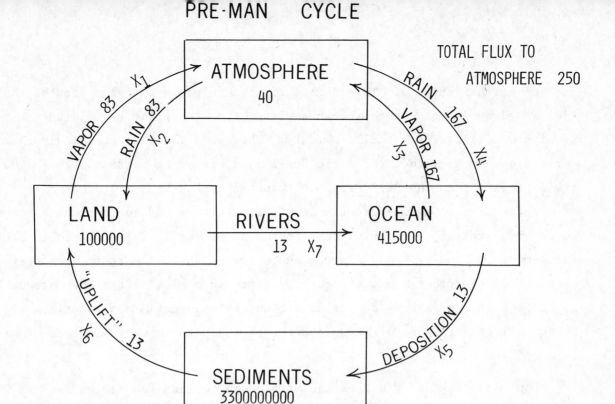

# PRE-MAN CYCLE

FIGURE 37. MODEL OF PRE-MAN CYCLE OF MERCURY.
RESERVOIR MASSES IN UNITS OF $10^8$ G;
FLUXES (X's) IN UNITS OF $10^8$ G/YR.

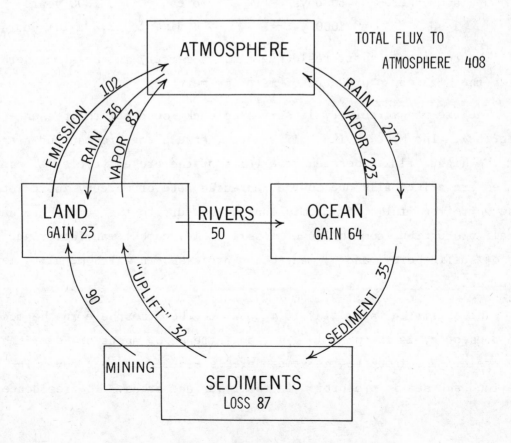

# PRESENT CYCLE

FIGURE 38. MODEL OF PRESENT-DAY CYCLE OF MERCURY.
RESERVOIR MASSES IN UNITS OF $10^8$ G;
FLUXES IN UNITS OF $10^8$ G/YR.

and the earth's surface. The flux of $250 \times 10^8$ g/yr is 20 times greater than that involved in land-stream-ocean-sediment transfer. The underlying cause is high volatility of metallic mercury. This volatility is strikingly demonstrated by the fact that the ratio Hg/Al in atmospheric particulates is 200 times greater than the Hg/Al ratio of crustal rocks (Fig. 35), the major source of particulates. Organo-mercuric substances in soil and sediments are known to degrade, releasing mercury vapor to the atmosphere. This process is bacterially mediated and probably also accounts for release of mercury from surface waters. On the other hand, the mercury fixed in shales, after they are buried beyond the zone of active bacterial action, is apparently non-volatile, for there is little change in the Hg content of shaly rocks over hundreds of millions of years.

The vapor is rained out of the atmosphere to land and sea surfaces. There is uncertainty about the details of the process; presumably mercury vapor dissolves in water droplets of clouds as elemental mercury. Figure 37 shows that vapor transport from the land and sea surfaces to the atmosphere before man was balanced by equal transfer of mercury in rain back to these surfaces.

The residence times of mercury in the four reservoirs were:

| | | | |
|---|---|---|---|
| atmosphere | 60 days | ocean | 32,000 years |
| land | 1000 years | sediments | $2.5 \times 10^8$ years. |

The relatively short residence time of mercury in the atmosphere re-emphasizes the high vapor pressure of metallic mercury.

The pre-man mercury cycle serves as background for the present-day cycle (Fig. 38). The most obvious differences between the pre-man and present cycles are the higher fluxes between reservoirs of the present cycle. These increased fluxes are principally due to the increased rate of mercury input into the land reservoir from mining, from emissions of mercury to the atmosphere during chlor-alkali production, combustion of fossil fuels, cement manufacturing, and roasting of sulfide ores, and by many other processes in which metallic mercury is involved.

The estimated vapor flux of mercury to the atmosphere in the model of the present cycle is $408 \times 10^8$ g/yr, an increase of about 60% over the pre-man cycle. The input to the atmosphere is assumed to be removed in rain over the land and sea in proportion to their surface areas. The residence time

Table 17

Estimates of Mercury Concentrations in the Environment

(in ppm)

| | |
|---|---|
| Rocks | 0.05-0.20 |
|   Shale | 0.15 |
|   Sandstone | 0.075     average ~ 130 |
|   Limestone | 0.075 |
|   Granite | 0.08 |
|   Basalt | 0.08 |
| Soils | 0.05 |
| Coal | ~1 |
| Oil | ~1 |
| Atmosphere | |
|   Total | $1 \times 10^{-6}$ |
|   Particulates* | 0.15 |
| Ocean | $3\text{-}15 \times 10^{-5}$ |
| Streams | |
|   Dissolved | $7 \times 10^{-5}$ |
|   Particulate* | 0.15 |
| Organisms | variable, up to ppm |
|   Human blood | 0.03-0.05 |
|   Marine skeletons | 0.40 |
|   Land biota | 0.5 |
|   Oceanic biota | 0.12 |
| Humus | 2-10 |

*The concentrations given are the Hg concentrations in the particulate material, not the concentration of Hg in the atmosphere or in streams obtained by dividing total atmosphere or stream mass by the total Hg in particulates.

---

of 36 days for mercury in the atmosphere justifies this assumption. Rain falling on the ocean results in a net input of mercury into the ocean surface of $49 \times 10^8$ g/yr; this value plus the addition of Hg to oceans via streams ($50 \times 10^8$ g/yr) minus the sedimentation rate ($35 \times 10^8$ g/yr) gives an accumulation of mercury in the ocean of $64 \times 10^8$ g/yr. Although some of this mercury enters living biota, most of it is dissolved in seawater. Figure 39 shows one prediction of possible future changes in the accumulation of mercury in seawater, if we continue to use mercury at a linearly increasing rate. In the year 2000, mercury concentration in the oceanic surface layer would have increased by 30%, not enough to harm plankton.

Several factors justify this conclusion. The total mass of mercury in living oceanic biota today is only $7 \times 10^8$ g, and the mercury concentrations

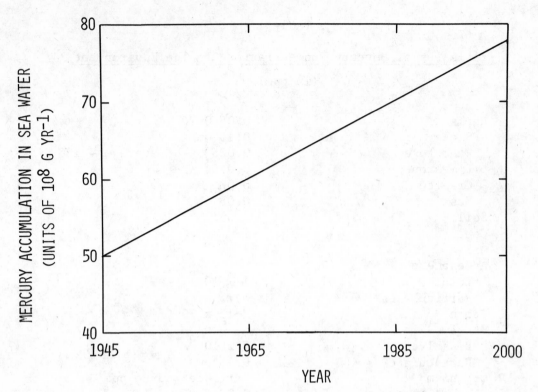

FIGURE 39.  PREDICTION OF POSSIBLE FUTURE CHANGES IN
ACCUMULATION OF MERCURY IN THE SURFACE
LAYER (UPPER 100 METERS) OF THE OCEANS.

in the organic and skeletal phases of the biota are only 0.12 and 0.40 ppm,
respectively.  The uptake of mercury in the photic zone by the biota is about
$58 \times 10^8$ g Hg/yr.  It would be expected that if a large part of the oceanic gain
of mercury each year entered the biota the total mass of mercury and its concentration in the biota would be greater than observed.  The calculated gain
in the ocean exceeds the biotic uptake rate.  A simple calculation further
justifies the conclusion:  assuming an ocean surface area of $360 \times 10^{16}$ cm$^2$
and a mixed layer depth of the ocean of 200 m, the rate of addition of mercury to this layer would result in an increase of concentration of about
0.0001g Hg/liter/yr.  This increase in total mercury is large enough to increase elemental mercury (Hg°) sufficiently in seawater to account for the
increased rate of evasion of mercury vapor from the sea surface today as
compared to pre-man time ($223 \times 10^8$ vs. $167 \times 10^8$ g/yr).

It is interesting to compare further the present total flux of mercury vapor to the atmosphere with the pre-man flux. The present rate is about 1.6 times the pre-man rate. Thus, we would predict that concentrations of mercury in glacial ice formed during recent time would be about twice as great as those of the past. Weiss, et al.[*] observed that water frozen into the Greenland ice sheet prior to 1952 contained $0.06 \pm 0.02$ ppb Hg, whereas the water deposited as ice from 1952 to 1965 contained $0.13 \pm 0.05$ ppb Hg, an observation in harmony with the conclusions of the preceding discussion.

The net result of mining and utilization of mercury on land and the increased mercury content of rain over land is a calculated increase of about 0.22% in the mercury content of the land surface (soil). Such an increase suggests that the degassing rate of the land surface during recent times may be greater than that prior to man's interference with the mercury cycle.

Mining and mercury utilization by man have increased the mercury content of rivers about 4 times. Today's rivers carry about equal masses of mercury in dissolved and solid form to the oceans. This relation is difficult to understand because, on the average, suspended sediment in rivers contains 3-4 times more mercury than the water. The problem may be that analyses of dissolved mercury in water include very fine suspended matter and mercury-organic complexes.

In summary, man's contributions to the mercury cycle rival the natural fluxes. The major pathways affected are from land to atmosphere to ocean and from land to ocean. It is likely that mercury is currently being stored on land and in the ocean. Mercury occurs in oceanic biota but it is not likely that on a global basis there has been a significant increase in their mercury content. The effects of the increase in atmospheric mercury on air-breathing organisms may be greater than effects on natural waters.

Toxic Concentrations

It should be clear from the low concentrations of mercury throughout most of the surface environment of the earth that unusual cricumstances are related to recent incidences of serious mercury poisoning. The inorganic compounds of mercury are toxic, but not remarkably so. Mercury vapor causes irritation and

---

[*] Weiss, H.V., Koide, M., and Goldberg, E.D., 1971. Mercury in a Greenland Ice Sheet: Evidence of Recent Input by Man, Science 174, 692-694.

destruction of lung tissues; miners often get tremors, irritation of gums, etc., but toxic levels are found only in mines or in areas of industrial use. Mercury used to preserve animal hides employed in manufacturing of hats in the 1800's led to 'hatter's disease', the syndrome that possessed the Mad Hatter in "Alice Through the Looking Glass".

Liquid mercury has an appreciable vapor pressure, so that spills from thermometers, barometers, etc., from which the tiny droplets are not cleaned up, can seriously contaminate a room. Inorganic mercury has a residence time in the human body of 2-6 months, so cumulative poisoning is quite possible. However, mercury compounds have long been used in medication, both internal and external, and surprisingly large amounts can be ingested without severe effects (one patient received a total of 78 grams without expiring!).

The now classic Minimata Bay disaster in Japan, in which many people died, therefore came as a surprise and a shock. Many of the inhabitants of the area, whose diet included lots of fish and shellfish, suffered from a variety of symptoms, including muscular weakness, loss of vision, brain injury, paralysis, coma, and in all too many instances, death. Investigation finally traced out the sequence of events responsible. Methyl mercury compounds were discharged directly into the Bay by a chemical factory. The methyl mercury was concentrated by microorganisms and algae and further concentrated by the fish that grazed on the organisms, and still further concentrated in fish that ate other fish. Some of the fish at the top of the food chain were found to contain as much as 50 ppm mercury as methyl mercury (50 times the level currently accepted in the U.S.).

Methyl mercury, as a toxic substance, behaves very differently from inorganic mercury compounds. It attacks nerve cells selectively, and is dangerous at very low levels, as compared to inorganic compounds. Thus, although the concentrations of inorganic mercury compounds in the waters and sediments of Minimata Bay were not apparently dangerous, the concentration of mercury in the food chain and the direct intorduction into the water of methyl mercury, created a disastrous situation.

Subsequent research has shown that several types of fish at the top of the food chain, even if not exposed to any man-induced mercury additions, contain as much as several ppm methyl mercury. Tuna and swordfish are two of the chief species involved. Initial discovery of high mercury levels in such fish raised the spectre of worldwide contamination by man's utiliaation of mercury; eventually it was shown, by chemical analyses of fish from museums, collected prior to important industrial mercury production, that the situation may be a natural one.

As it now stands, it is clear that inorganic mercury compounds cannot be discharged into environmental situations where methylation and food chain concentration are possible. In most cases there has been rapid compliance of industry to this requirement. Also, it is clear that certain fish must not be eaten to excess, and that catches, even from unpolluted areas, must be monitored for Hg content. Finally, the eating of most fish, unless carried to excess, is safe. The highest mercury content reached by fish under 'natural' conditions is about 1 ppm, but it is probably safe to have swordfish or tuna from time to time, even at the higher levels.

In addition to the methyl mercury in fish, it has been discovered that grain treated with methyl mercury as a preservative has caused serious poisoning. Extensive analyses of most staple foods show them to be under the current accepted limit of 0.05 ppm, and there has been little change in 30 years. Also, most all streams and lakes (and their accompanying sediments) in the U.S. have levels of mercury below those that can cause difficulties, even by food chain concentration.

## Uses of Mercury

Care must be exercised in the disposal of mercury, but its uses ramify throughout our current pattern of life. It is used in thermometers, barometers, in paint, pesticides, fungicides, antiseptics, floor wax, furniture polish, fabric softeners, air conditioning filters, and as an anti-mildew agent. Industrially its greatest utilization is as electrodes in various electrolyte processes, and it has been leakage from some of these large scale processes that has caused dangerous or incipiently dangerous conditions, where methyl mercury could form and be concentrated by organisms.

## Conclusions

The case history of mercury is likely to be representative of that of other effects of man's additions to natural cycles. First, there are serious consequences of additions; second, there is intensive investigation of the reasons for these consequences; third, controls are developed and enforced to alleviate the effect of man's additions; finally there exists the capability of determining whether the additions will have controllable and temporary consequences, or whether permanent damage will have been done to future ecosystems.

## Lead

### Cycling

Lead, like mercury, has been a widely publicized pollutant. Also, like mercury, increased knowledge of its behavior has produced information that has already been effective in reducing the harmful effects that originally brought attention to lead as a pollutant.

The original outcry came from the realization that tetraethyl lead, added to gasoline to increase engine efficiency, was being released to the atmosphere in large quantities, and for the first time in the history of the earth, lead was being cycled as a volatile component rather than as one confined to water transport (excluding some transport of lead as a particulate in the atmosphere).

Figure 40 shows an estimate of some of the current fluxes and reservoirs of the cycle of lead. Igneous rocks contain about 10 ppm lead. Total lead in the sedimentary rock reservoir is estimated at $4 \times 10^{19}$ grams, based on an average value of 16 ppm. Weathering today accounts for about $240 \times 10^{14}$ g of rock per year; thus, the lead content of materials eroded is $16 \times 10^{-6} \times 240 \times 10^{14}$ g/yr = $384 \times 10^{9}$ g/yr.

The lead mined each year is estimated at $3100 \times 10^{9}$ g/yr, illustrating a now familiar situation; the lead mined, which is taken only from a few places of relatively very high concentration (several %), exceeds the lead deposited and eroded in the natural cycle by a large ratio. 'Economic' lead is being depleted at a high rate indeed, and is a non-renewable resource.

The rate at which lead mined is being fed back into the oceans (its eventual sink) is completely unknown. Lead has a wide variety of industrial uses, with the greatest single demand being for storage batteries. The uses for lead are changing rapidly as its toxic character becomes more widely recognized. It was used widely in paints and plumbing until its poisonous effects were realized. Much lead goes into bullets and shotgun pellets; appreciable quantities are used as shields against radioactivity. A recent, still small, but increasing use is for weights in belts of scuba divers. At any

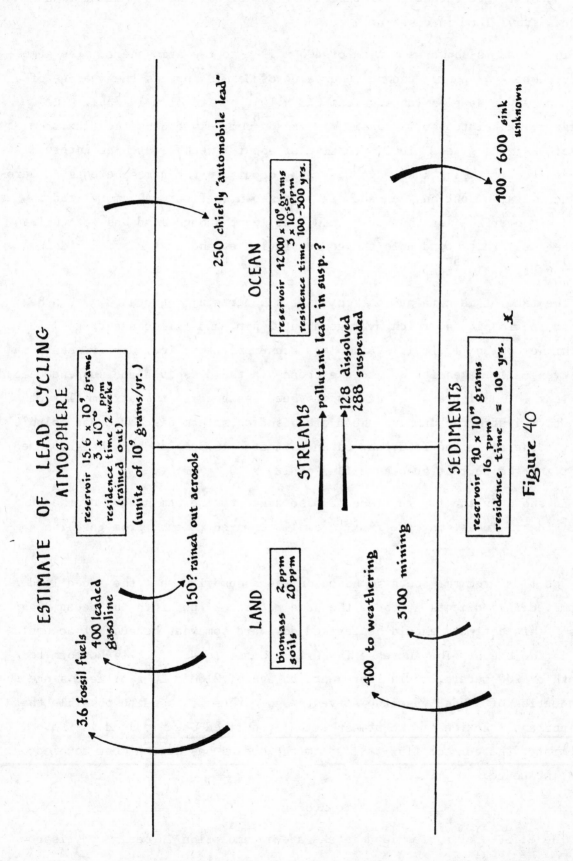

ESTIMATE OF LEAD CYCLING

ATMOSPHERE

reservoir 15.6 × 10⁹ grams
3 × 10⁻⁶ ppm
residence time 2 weeks
(rained out)

(units of 10⁹ grams/yr.)

3.6 fossil fuels

400 leaded gasoline

150? rained out aerosols

LAND

biomass 2 ppm
soils 20 ppm

400 to weathering

3100 mining

STREAMS

pollutant lead in susp. ?

128 dissolved
288 suspended

OCEAN

reservoir 12000 × 10⁹ grams
3 × 10⁻⁵ ppm
residence time 100-300 yrs.

250 chiefly "automobile lead"

100 - 600 sink
unknown

SEDIMENTS

reservoir 90 × 10¹⁵ grams
16 ppm
residence time = 10⁸ yrs.

Figure 40

rate, according to recent estimates, about 10% of all the lead mined is used as tetra-ethyl lead in gasoline.

This lead is shown as a flux of $400 \times 10^9$ g to the atmosphere. The atmospheric content is poorly known; the value of $3 \times 10^{-6}$ ppm is the average of a large number of samples ranging from $0.3 \times 10^{-6}$ ppm to $11 \times 10^{-6}$ ppm. The reservoir, calculated on the basis of $3 \times 10^{-6}$ ppm, divided by the flux from tetraethyl lead, gives a residence time of about 2 weeks. Such an interval is consistent with the knowledge that lead occurs in the atmosphere as an aerosol and is rained out on land and sea. As shown in Figure 40, more lead is rained out over the sea than over land; at least enough lead has gotten into surface ocean water to double the concentration at the surface from $3 \times 10^{-5}$ ppm to 6 or $7 \times 10^{-5}$ ppm.

The fate of lead rained out on land is uncertain; undoubtedly much has been stored in plants, which average about 2 ppm, and will concentrate lead to much higher values if it is available. Presumably this increased vegetational lead eventually makes its way to the ocean via the dissolved and suspended load of streams. In Figure 40 no number has been assigned to the stream flux of tetraethyl lead; the numbers shown are based on stream discharge and soluble lead for the dissolved load, and on an estimated 16 ppm lead in the suspended load times the current suspended flux of $180 \times 10^{14}$ g/yr.

As shown in the figure, lead is also added to the atmosphere by the burning of fossil fuels, but the flux is small compared to that from gasoline combustion.

The flux from the ocean is based on the assumption that the added lead is removed in sediments at about the same rate it is put into the ocean; obviously this estimate should be revised downward somewhat because of accumulation in the ocean. The increase in surface water lead, if it is uniform to a depth of 200 meters, would represent storage of $2160 \times 10^9$ g of lead since the introduction of leaded gasoline 45 years ago; thus it is quite possible that an important fraction of the atmospherically derived oceanic lead is stored. In summary, the current flux of lead to the oceans has been almost doubled by man's influence.

## Toxicology

The effects of a high lead intake are accumulation in bones and tissues. Among the symptoms are mental deficiency and behavior problems. On the cellular level, lead inhibits enzyme activities. Acute poisoning can cause anemia, kidney impairment, and brain damage.

Most lead poisoning stems from eating lead, not from breathing it, even in areas of high traffic density. In the U.S., average intake in diet is 0.3 milligrams/day, and less than 0.03 mg/day is absorbed. Respiratory intake ranges between 0.005 and 0.05 mg/day. The upper limit of lead concentration for water supplies has been placed at 0.05 ppm. For U.S. cities, the maximum in over 100 cities studied was 0.07 ppm; the average only 0.004 ppm. Lead metabolism in the normal person is discussed in Chapter 12.

A tentative limit for atmospheric lead has been set at about 0.007 ppm. A ten week study of the air at 45th Street in New York City showed about 0.008 ppm as an average; another year-long study on Broadway, New York City, showed about the same level. One concludes that direct human damage from inhalation is probably minimal.

A major effect of lead released to the atmosphere is to enrich the plants and soils adjacent to the source. The plants in turn can be eaten by animals with resulting serious toxic effects. Soils adjacent to highways have been found to contain hundreds of ppm lead; the concentration falls off logarithmically, but above-normal levels can be detected up to 10's of miles away.

Twigs of trees adjacent to a Vancouver highway contained 3000 ppm lead in their ash; this value is about 100 times the unpolluted value. Plants apparently can take up 'automobile lead' more efficiently than that from lead-containing minerals.

As stated before, lead poisoning is chiefly a dietary problem. By far the leading cause is children eating peeling lead-based paints in old buildings. There are many other sources: lead glazes on earthenware used for cooking or drinking vessels have led to severe lead poisoning; use of lead pipes, a former source of high lead content in drinking water, is essentially completely discontinued in the U.S. The early symptoms of lead poisoning are difficult to detect, being rather non-specific, but today treatment of even fairly advanced cases is successful. The patient is given an organic complexing agent which solubilizes the deposited lead and permits it to be excreted.

## Summary

Lead additions to the environment by man now rival the amounts naturally cycled. Lead additions to the atmosphere, chiefly by automobiles, do not seem

131

to be a direct threat by poisoning due to inhalation, but they have secondary effects on land, such as increasing markedly the lead levels in vegetation. The soluble lead content in ocean surface waters has doubled; the consequences are unknown. Most dietary sources of lead have been or are being eliminated.

## Manganese

The global cycles of manganese as shown in Figures 41 and 42 illustrate the differences we can expect in trace element behavior owing to differences in relative volatilities of the elements. As potential pollutants, those elements with high to moderate volatilities are most likely to have greatest impact on natural global cycles because they can travel long distances as vapors; low volatility elements, like Mn, tend to settle out near sources and be of more concern in local or regional environmental problems, although as fine particulate material, they too, may be distributed worldwide.

There are several striking differences between the cycles of mercury, a volatile element, and manganese, a non-volatile element. First, the major exchange of manganese between the atmosphere and the pre-man earth surface was due to continental dust being swept into the atmosphere by winds and then falling back onto the earth's surface. Today this dust flux is augmented by manganese emitted to the atmosphere in particulate form by industrial activities (e.g., steel manufacturing). Bacterial processes or high temperature release and then condensation processes do not appear to be important for manganese; that is, no volatile manganese species is important in the element's global cycle.

Second, the total river flux of manganese to the ocean today is nearly three times the pre-man flux. This increase represents principally an increase in the denudation rate of the land's surface from about $100 \times 10^{14}$ g/$yr^{-1}$ pre man to today's rate of about $225 \times 10^{14}$ g/$yr^{-1}$. Because this increase in denudation reflects an increase in the suspended load of rivers from deforestation and agricultural activities, and because manganese is concentrated in the ferric oxide coatings on suspended material and in the suspended particles, the land to ocean manganese flux is higher today than in the past.

Third, man's activities have changed the global cycle of manganese. Manganese in particulate emissions from industrial activities rivals the natural input of continental dust to the atmosphere ($30 \times 10^{10}$ g vs. $43 \times 10^{10}$ g). Most

PRESENT DAY CYCLE OF MANGANESE

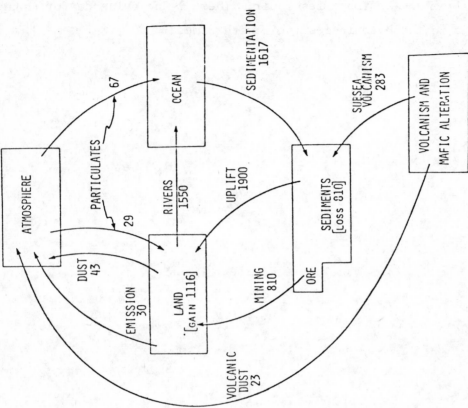

FIGURE 42. PRESENT-DAY CYCLE OF MANGANESE.
FLUXES ARE IN UNITS OF $10^{10}$ G/YR.;
RESERVOIR MASSES IN UNITS OF $10^{10}$ G.

PRE-MAN CYCLE OF MANGANESE

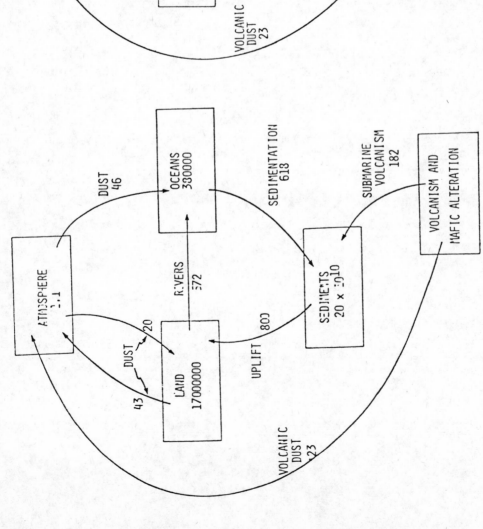

FIGURE 41. PRE-MAN CYCLE OF MANGANESE.
FLUXES ARE IN UNITS OF $10^{10}$ G/YR.;
RESERVOIR MASSES IN UNITS OF $10^{10}$ G.

particulate Mn probably falls out of atmosphere near industrial sources. The mining of manganese ore has resulted in a net gain for the land reservoir and a net loss from the sediment reservoir. There is no evidence for change of dissolved Mn in the oceanic reservoir with time.

Study Questions

1. Cadmium concentration in urban particulates is 2 ng/m$^3$. Calculate its worldwide emission to the atmosphere by industrial practices.

2. If cadmium had a worldwide atmospheric concentration of 0.1 ng/m$^3$, what would be its total atmospheric rainout?

3. From questions 1 and 2, what is the interference factor for cadmium in the atmosphere?

4. From examination of Figure 35, what elements other than mercury might you expect to be present in the atmosphere in a form other than particulate?

5. What are some of the beneficial and detrimental aspects of trace elements?

6. From Figure 36, what elements would you be most concerned about in terms of pollution of public water supplies? Why?

7. Total mercury in sedimentary rocks is estimated at 330 x 10$^{15}$ g.
   (a) What is the number of moles of Hg in sedimentary rocks?
   (b) If the total mass of sedimentary rocks is 26,000 x 10$^{20}$ g, what is their content in ppm of Hg?

8. Figure 38 shows the total amount of Hg carried annually to the oceans by streams as 50 x 10$^8$ g. Average stream concentration (Table 17) is 0.07 micrograms per liter. If annual stream flow is 0.32 x 10$^{20}$ g/yr, what percentage of total stream Hg is dissolved?

9. Atmospheric concentrations of Hg near the ground have been used successfully in prospecting for mercury deposits, but measurements of Pb concentrations have been unsuccessful in the search for lead. What basic difference in the properties of Hg and Pb compounds accounts for these findings?

10. The concentration of lead in streams (Table 5) is given as 3 micrograms/liter, that of mercury as 0.07 micrograms/liter. What does this suggest concerning the solubility of mercury minerals?

11. If mercury in foods averages about 0.02 ppm, and we eat about 1 kg/day food, what is the daily intake of Hg in micrograms? Compare this with the daily ingestion of lead.

12. Where is most of the mercury released into streams from industry now stored? Do you regard this reservoir as a future peril? Explain.

13. In principle, would you worry more about mercury poisoning from eating shark meat or from eating one of the fish that gets its food by grazing on reefs? Document your argument.

14. Calculate from Figure 42 the interference factor for Mn in the atmosphere.

15. Why is trace element toxicology such a difficult field, especially when it comes to putting safe upper limits on permissible concentrations?

16. As has been emphasized, it is the production of methyl mercury by bacterial activity that makes Hg more hazardous than Pb as a toxic element, especially because of its concentration in the food chain. As an environmental researcher, what would be your line of attack on other trace elements, such as arsenic, antimony, cadmium, etc.?

    This is a discussion question for class and instructor. Some lines of attack are:

    1. Study physical-chemical properties of element.
    2. Determine theoretical abundances in waters, etc., from chemical (thermodynamic) data.
    3. Study bacterial-mediated reactions and rates.
    4. Evaluate routes of uptake in the food chain and possible concentrations.
    5. Obtain data on concentrations in natural systems.
    6. Evaluate toxicology in experiments.
    7. Construct local and global models.
    8. Assess impact on environment.

Chapter 11


SYNTHETIC ORGANICS, PETROLEUM AND PARTICULATES


## Synthetic Organics

Many different kinds of synthetic organic compounds foreign to the natural system enter the environment each year from agricultural, sanitation and industrial practices. Those chemicals of most immediate interest have three things in common: they are produced in large quantities, are relatively stable, and are toxic. Some examples are DDT, PCB's (polychlorinated biphenyls), chlordane, dieldrin, freon, and dry-cleaning solvents such as carbon tetrachloride. Synthetic compounds have cycles and undergo reactions in the environment similar to those of natural materials. As an example of the case histories of these synthetic organics, DDT will be discussed here; however, it should be recognized that some statements concerning DDT's potential for environmental damage are still equivocal.

### DDT Toxicology and Use

DDT is an organic compound of C, H and Cl and is slightly volatile, virtually insoluble in water but moderately soluble in organic solvents, and particularly soluble in fatty tissue. It degrades slowly to residues such as toxic DDE and non-toxic DDD. The time required for degradation of half of a given amount of DDT (half-life) is 10-20 years. The following abbreviations are for the compounds:

DDT = (1,1,1-trichloro-2,2-bis (p-chlorphenyl)-ethane

DDE = (1,1-dichloro-2,2-bis (p-chlorphenyl)-ethylene

DDD = (1,1-dichloro-2,2-bis (p-chlorphenyl)-ethane.

DDT is a general poison that affects essentially all organisms. In very large doses it is lethal to man and the larger mammals. It attacks the central nervous system and also has a broad range of other pathological effects on organisms (Table 18). Birds are particularly susceptible; effects on them include eggshell thinning, impaired bone formation, and reduction of carbonic anhydrase activity. Carbonic anhydrase is an enzyme that catalyzes reactions of dissolved carbon dioxide species, and helps in the formation of $CaCO_3$ shells.

Table 18

Physiological Effects of DDT Residues and PCB's on Organisms in Laboratory Tests

| Organism | Concentration | Result |
|---|---|---|
| Phytoplankton | few ppb DDT in water | inhibited photosynthesis |
| Phytoplankton | 7 ppb PCB's in water | inhibited photosynthesis |
| Fresh-water algae (blue-green) | 800 ppb DDT in NaCl solution | inhibition of NaCl tolerance |
| Kestrel Hawk | fed a diet containing 3 ppm DDE; 3-22 ppm DDE in egg | egg shell thinning |
| Quail | fed a diet containing 100 ppm DDT or DDE; 48-244 ppm DDT residues in egg | reduction of carbonic anhydrase activity |
| Ringdove | fed a diet with 10 ppm DDT; 0.13-103 ppm DDT residues in brain | delay in egg-laying; decrease in medullary calcium and eggshell weight |
| Ringdove | injected with 150 ppm DDE; 17 ppm DDE in egg | reduction of eggshell weight and carbonic anhydrase inhibition |
| Sparrow Hawk | fed a diet containing 0.8 ppm dieldrin and 5 ppm DDT | reproductive failure |
| Pigeon | fed a diet containing 100 ppm DDE | reduction in medullary bone formation |
| Mallard | fed a diet with 10-40 ppm DDE | egg-shell thinning |
| Brine Shrimp | exposure to DDT in ppm range in solution | mortality |
| Copepods | $5-10 \times 10^{-6}$ ppm DDT in water | mortality; blockage of adult development |
| Crabs | exposure to DDT in ppm range in water | mortality |
| Oysters | 0.1 ppb DDT in water | inhibits growth |
| Trout (freshwater) | 5 ppm DDT in eggs | inhibits development of fish fry |

Eggshell thinning is probably due to interference of DDE with calcium transport in the shell gland, although earlier hypotheses emphasized that DDT interfered with the production of estrogen, a sex hormone governing calcium metabolism. Whichever the case, a direct correlation has been found between eggshell thickness and DDE concentrations in the eggs of several species of birds; a marked decrease in shell thickness was found after uptake of small amounts of DDE. Decrease in shell thickness results in egg breakage during laying or incubation.

DDT as a pesticide has killed billions of insects that destroy crops and carry and transmit such diseases as malaria (mosquitoes), typhus fever (lice), and plague (fleas). Use of this insecticide has improved the economic, social and health status of underdeveloped countries. (In 1953, DDT was credited with having saved 5 million lives since its use began and curtailing 100 million illnesses, e.g., curbed malaria in Italy, typhus fever in Germany after WW II, and plague in Dakar in 1944.)

Because DDT is volatile, soluble in fats, and slow to degrade, it has been dispersed across the earth's surface. The degree of dispersal can readily be seen by examination of Table 19. Oceanic birds and marine fishes from localities distributed throughout much of the northern hemisphere contain significant amounts of DDT. Many concentrations shown in Table 19 are within the range of concentrations found in organisms that showed adverse physiological effects of DDT in laboratory experiments (Table 18). In the southern hemisphere, penguins as far south as Antarctica contain DDT.

DDT has been responsible on the local scale for fish kills and presumably for decline of bird populations (e.g., osprey, falcon, Bermuda petrel). It was shown to reduce photosynthesis in phytoplankton in laboratory experiments (DDT in phytoplankton from Monterey Bay, California, is three times greater now than in 1955). Residues have been found in numerous organisms. Some breast-fed babies in the U.S. were discovered to have concentrations twice the amount deemed safe by the World Health Organization (0.010 mg DDT/day/kg body weight).

Because of its widespread dispersal and toxicity, the Environmental Protection Agency has banned the use of DDT in the U.S., except in cases of public health emergencies and in application to certain crops (onions in the Pacific northwest, sweet potatoes in storage, and peppers along Chesapeake Bay). Its continued use in other countries is likely for at least several more years.

139

Table 19

DDT and PCB Residues in Oceanic Birds and Fishes

| Species | Locality | DDT (wet wt., ppm) | PCB (wet wt., ppm) |
|---|---|---|---|
| Shearwater | Mexico | 3.0 | 0.4 |
| " | California | 11.3 | 1.1 |
| " | New Brunswick | 40.9 | 52.6 |
| " | California | 32.0 | 2.1 |
| " | New Brunswick | 70.9 | 104.3 |
| Petrel | Bermuda | 6.4 | - |
| " | California | 66.0 | 24.0 |
| " | Mexico | 9.2 | 1.0 |
| " | Mexico | 3.2 | 0.35 |
| " | Baja California | 953 | 351 |
| " | New Brunswick | 164 | 192 |
| " | New Brunswick | 199 | 697 |
| Anchovy | San Francisco Bay | 0.33-0.59 | - |
| " | Monterey | 0.90 | - |
| " | Morro Bay | 0.74 | - |
| " | Port Hueneme | 3.04 | - |
| " | Los Angeles | 14.0 | 1.0 |
| English sole | San Francisco Bay | 0.19-0.55 | 0.05-0.11 |
| " | Monterey | 0.76 | 0.04 |
| Shiner perch | San Francisco Bay | 1-1.4 | 0.4-1.2 |
| Jack Mackerel | Channel Islands | 0.56 | 0.02 |
| Hake | Puget Sound | 0.18 | 0.16 |
| " | Channel Islands | 1.8 | 0.12 |
| Bluefin tuna | Mexico | 0.22-0.56 | 0.04 |
| Yellowfin tuna | Galapagos Islands | 0.07 | - |
| " | Central America | 0.62 | 0.04 |
| Herring | Baltic Sea | 0.68 | 0.27 |
| Plaice | Baltic Sea | 0.018 | 0.017 |
| Cod | Baltic Sea | 0.063 | 0.033 |
| Salmon | Baltic Sea | 3.4 | 0.30 |

Adapted from Risebrough, R. W., 1971, Chlorinated Hydrocarbons: in Impingement of Man on the Oceans, D. W. Hood, ed., Wiley-Interscience, New York, 259-286.

## DDT Global Model

Estimates of DDT residue concentrations in the environment are shown in Table 20. The estimates, although reasonable, should be viewed with caution; few analyses of DDT residues in seawater or the atmosphere are available. Only recently it was found that much of the DDT in the atmosphere may be in the vapor phase -- thus previous estimates of atmospheric DDT concentrations may be in error because the analytical methods did not include measurement of DDT in the gaseous state.

Nevertheless, estimates of residues have been used to build models of the global cycle of DDT. One such tentative model is shown in Figure 43. A more quantitative model was presented by Woodwell, et al.[*], in an article in Science in 1971. A major criticism of these models by some investigators is the conclusion that there is major atmospheric transport of DDT to the oceans. Also, recent scientific findings that need further confirmation would modify the present models. They are (1) studies of the distributions of DDT and PCB's in the atmosphere and surface seawater of the Sargasso Sea area yield a residence time of DDT in the local atmosphere of only about 50 days, as contrasted to the global value of 3-4 years previously used; (2) experiments have shown that DDT vapor can be converted to PCB's by ultraviolet light of the same wavelengths found in sunlight in the troposphere. This finding may account for the widespread distribution of PCB's over the earth's surface (Table 19), but also would necessitate another sink for DDT in any model of its global distribution.

The simplified model of DDT in Figure 43 is based on an average world use of DDT of $8 \times 10^{10}$ g/yr for the past 20 years. This assumption is not strictly true since DDT production began in the early '40's, rose exponentially to the late '60's, and appears to have declined slightly in recent years. The figure, however, demonstrates clearly the conclusion that the major route of transfer of DDT and its residues to the oceans is through the atmosphere as a vapor or adsorbed to dust particles. About 25% of the DDT added to the land annually enters the oceans via the atmosphere whereas only 0.5% is transported to the oceans by streams. The gains in the reservoirs, land ($5 \times 10^{10}$ g/yr), ocean ($2 \times 10^{10}$ g/yr), atmosphere ($1 \times 10^{10}$ g/yr) and sediments ($0.08 \times 10^{10}$ g/yr),

---

[*] Woodwell, G.M., Craig, P.P., and Johnson, H.A., 1971. DDT in the biosphere: Where does it go?, Science 174, 1101-1107.

## Table 20

DDT Residue Concentrations in the Environment (ppm)

|  | ppm |
|---|---|
| **Soils** | |
| agricultural | 0.11 |
| non-agricultural | 0.0003 |
| **Atmosphere** | |
| total | $1 \times 10^{-8}$ to $2 \times 10^{-3}$ |
| dust | 0.014 |
| **Rain** | $60 \times 10^{-6}$ |
| **Terrestrial biota** | |
| plants | 0.0001 |
| animals | 1 |
| **Marine biota** | |
| plankton | 0.01 |
| fish | 1 |
| mammals | variable, <1 to >1000 |
| **Oceanic mixed layer** | $6 \times 10^{-8}$ to $10 \times 10^{-6}$ |
| **Rivers** | |
| water | $3 \times 10^{-6}$ |
| suspended sediments | 0.014 |

------------------------------------------------------------

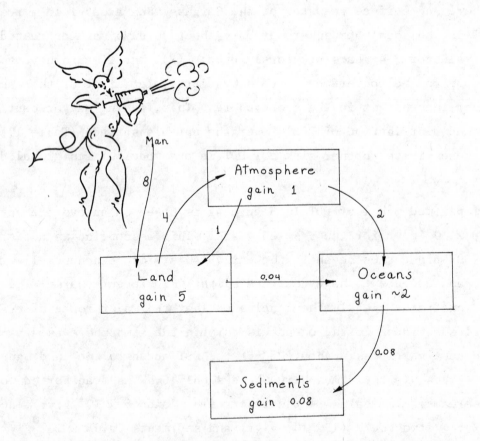

FIGURE 43. SIMPLIFIED DDT CYCLE.
FLUXES AND RESERVOIR GAINS IN UNITS OF $10^{10}$ G/YR.
DDT INCLUDES DDE AND DDD AND OTHER RESIDUES.

integrated over the 20 years in which DDT production has averaged about $8 \times 10^{10}$ g/yr, represents a total accumulation of DDT and its residues in these reservoirs of $1.6 \times 10^{12}$ g. This value is reasonably close to the total integrated world production from 1940 to 1974 of $2 \times 10^{12}$ g.

Woodwell, et al., in their detailed global model of DDT attempted to predict the fate of DDT in the troposphere and oceanic mixed layer, assuming two different initial conditions. Their results are shown in Figure 44. Of interest is the prediction that if DDT production in the U.S. declined to zero in 1974, DDT concentrations in the troposphere and oceanic mixed layer would decline to 1950 values by the years 1985 and 1995, respectively. If there were increasing use of DDT through 1980, DDT residue concentrations would continue to rise in the troposphere and oceanic mixed layer but the rise would be less pronounced than prior to 1965.

Most models of DDT also show that the oceanic and terrestrial biota contain only a small amount of the total DDT released to the environment. Most of it remains on land or accumulates in the ocean and atmosphere. Furthermore, DDT increases in the atmosphere and oceans follow soon after its addition to the earth's surface environment.

The DDT problem is an excellent example of what can happen when a toxic substance is used widely without assessment of its hazards. It is fortunate that little DDT actually has entered biota (perhaps a total of 1/30 of one year's production in the mid-1960's), considering the effects of a small amount on bird and fish populations. Even today our knowledge of the DDT cycle is meager; we are still not sure where it all goes. Perhaps in the future the toxicology and distribution patterns of other synthetic organics will be assessed before they are released to the environment.

## Petroleum

Approximately $4 \times 10^{14}$ g of petroleum (Table 21) are transported across the sea annually (increasing at 4%/yr); of this amount, about $2 \times 10^{12}$ g are lost to the oceans annually during marine transportation. This mass exceeds that of hydrocarbons produced naturally each year by organisms in the ocean. A single Torrey Canyon incident delivers $10 \times 10^{10}$ g of petroleum to the sea. The blowout of the Santa Barbara oil wells added a million gallons of petroleum to the ocean.

Table 21

Annual Fluxes of Petroleum and its Byproducts*

| Source | Flux (units of $10^{12}$ g/yr) |
|---|---|
| 1971 World oil production | 2500 |
| 1971 Oil transport by tanker | 1400 |
| Anthropogenic injections to marine environment | |
| Loss from offshore ore production | 0.08 |
| Marine transportation | 2.13 |
| Loss from coastal oil refineries | 0.2 |
| Industrial waste | 0.3 |
| Municipal waste | 0.3 |
| Urban runoff | 0.3 |
| River runoff[a] | 1.6 |
| Subtotal | 4.91 |
| Natural seeps | 0.6 |
| Atmospheric rainout[b] | 0.6 |
| Total | 6.11 |
| Torrey Canyon accident | 0.1 |
| Santa Barbara well blowout | 0.003-0.1 |
| Total vaporization of petroleum products from continents to atmosphere | 90 |

([a] In part natural; [b] in part man's additions)

Adapted from Massachusetts Institute of Technology.  Man's Impact on the Global Environment:  Report of a Study of Critical Environmental Problems, Cambridge, MIT Press, 1970, 319 pp;  National Academy of Sciences.  Petroleum in the Marine Environment.  Washington, National Academy of Sciences, 1975, 107 pp.

----------------------------------------------------------------------------

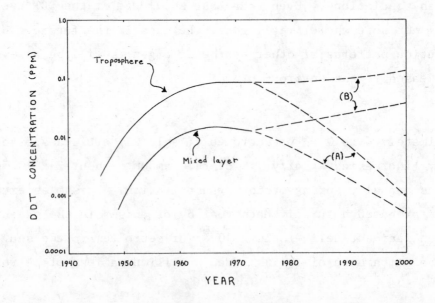

Figure 44.  DDT concentration in the troposphere and organic mixed layer predicted on assumption of (a) declining use of DDT through 1974 and (b) increasing use of DDT to 1980.  Adapted from Woodwell, et.al., (op. cit.).

## Behavior in Seawater

Light fractions evaporate and decompose in the atmosphere, some oil absorbs on particulate matter and sinks, some dissolves and some is oxidized by bacteria. Fuel oil hydrocarbons, however, appear to be persistent; some fractions exist in the marine environment probably for years or decades. Two years after a spill in Buzzards Bay, Massachusetts, the hydrocarbons were not completely degraded and were found in marsh and offshore sediments. Tar lumps have been observed in the open sea since around 1968 and reach concentrations of 2-40 mg/m$^2$ in the Sargasso Sea. These tar lumps accrete on beaches (such as in Bermuda); their chemical characteristics show their origin to be crude oil sludge from tanker washings. The life time of these lumps at sea is a year or more.

## Effect on Biosphere

(1) Most visible is death of birds owing to physical effects of the oil (flying difficulty, loss of buoyancy, etc.).

(2) Concentrates in lipid portion of organisms ('oily tasting fish'). Marine organisms contain 1 to 400 ppm petroleum hydrocarbons (wet weight). Some organisms apparently can discharge hydrocarbons readily, however, oil can destroy whole communities of organisms in large spills (e.g., Tampico Mara spill in 1957 off Baja California). At least 15 major oil spills have occurred since 1957. There is no evidence for food chain magnification -- direct uptake of hydrocarbons by organisms from water or injestion of sediment seems to be the rule.

(3) May interfere with chemically stimulated behavior (example: starfish being attracted to oyster prey because of release of small amounts of hydrocarbons by the oyster; may be true of lobster also).

(4) Some fractions may be carcinogenic (e.g., benzopyrene, a carcinogen, found in concentrations of 0.016 to 6 ppm in sediments of the French Mediterranean; also found in plankton).

(5) Aesthetic: oil on beaches, slicks from spills, etc.

## Particulates

The atmosphere has always contained a burden of dust derived principally from soils by wind. Man, through agricultural and industrial practices, has increased the dust flux to the atmosphere. Table 22 compares natural dust fluxes with those owing to man's activities. The following is a brief summary of some

145

Table 22

Fluxes of materials brought into the exogenic cycle
by man and by natural processes

| Material | Reservoir receiving material | Flux $10^{14}$ g/yr |
|---|---|---|
| **natural processes** | | |
| * sea salt | atmosphere | 1.0 |
| suspended river solids | ocean | 180.0 |
| dissolved river load | ocean | 39.0 |
| *continental rock and soil | atmosphere | 5.0 |
| mercury | ocean | 0.00018 |
| *volcanic debris | stratosphere | 0.036 |
| | atmosphere | 1.5 |
| lead | oceans | 0.0042 |
| **man** | | |
| *carbon and fly-ash from fossil fuel combustion | atmosphere | 0.25 |
| DDT residues | atmosphere | 0.0004 |
| | oceans via rivers | 0.000004 |
| mercury | atmosphere and rivers | 0.00032 |
| lead | atmosphere | 0.004 |

(*after Goldberg, E.D., 1971. Atmospheric dust, the sedimentary cycle and
man. Comments on Earth Sciences: Geophysics, 1, 117-132.)

------------------------------------------------------------------------

of the particles found in the atmosphere.

Dust: most important particle in atmosphere; derived from soils (1930's
'dust bowl'); total flux to atmosphere about 1 to 5 x $10^{14}$ g/yr.

Soot and fly-ash: second most important particle; finely divided carbon
clumped together (soot); fly-ash is colorless glass spheres. Both are derived from combustion processes and are estimated at 0.25 x $10^{14}$ g/yr. Soot
generally carries heavy hydrocarbons with it. Some of these hydrocarbons include compounds that are cancer-producing. In some urban communities fallout
of soot, dust and fly-ash may amount to 1 lb/yr/sq.ft.

Smog particulates: little known about them but they result from chemical
reactions in smog, are oily, and obscure visibility.

Re-suspended particulates: particles of newspaper, rubber, glass, etc.,
which are blown into the air, settle, and become re-suspended.

Talc: mineral particulate (Mg silicate), widespread, found in seawater, rain, air, glaciers; used as a filler and diluent for pesticides and other sprayed substances (i.e., crop dusting) and as propellant in aerosol cans.

Asbestos: little known, but in 1968, $7 \times 10^9$ g escaped to the atmosphere. Implicated in lung diseases of miners exposed to it.

Cellulose: found floating in ocean water in world oceans. Off eastern U.S. coast, much derived from toilet paper.

Major environmental concerns with respect to man-produced particulates are:

(1) They can be irritants leading to respiratory illnesses,

(2) They may carry adsorbed organics of a carcinogenic nature,

(3) In large concentrations in the atmosphere they may change earth's temperature; thought to be responsible for recent cooling trend of earth.

Study Questions

1. Give the basic arguments pro and con for continued use of DDT.

2. Why is DDT so especially harmful to birds like eagles, falcons, and pelicans?

3. DDT is known to slow the rate of photosynthesis in marine organisms. Would slowing of photosynthesis cause a significant drain on atmospheric oxygen?

4. If $8 \times 10^{10}$ g of DDT were added to the earth's surface in 1974, how much would be left in 2004? Assume half-life of DDT is 15 years.

5. What is the source reservoir of DDT found in ice and snow in Antarctica?

6. PCB's are a group of compounds with properties of non-flammability, high boiling temperature, temperature independent viscosity and water insolubility. They have a wide range of uses as cooling media, plastic softeners, and additives in ink, paints, etc. They are distributed throughout the environment at concentrations similar to DDT. In most organisms the PCB's have chemical structures that match very well those of commercial PCB brands. However, in seals and certain bottom-feeding organisms, the structures differ from those of commercial formulations of PCB's. Give a possible reason for this difference.

7. If the worldwide production of DDT ceased today and its global half-life is 15 years, what would be the residence time of DDT in the oceans 30 years from now?

8. Where in the oceans would you expect oil blobs to accumulate?

9. Into what two fractions does spilled oil split? Which do you regard as more dangerous? Why?

10. In 1970 approximately $10 \times 10^{14}$ g of petroleum were transported across the ocean of which 0.1% was lost to the ocean by spillage and leakage. At an annual increase of 4% in the tonnage of petroleum transported by ships, what will be the total loss of petroleum in 1975 owing to spillage and leakage, assuming 0.1% is lost?

11. What is the interference factor for soot and fly-ash in terms of the total annual transport of dust from continents to ocean?

12. If the soot and fly-ash carry 0.5 percent carcinogenic compounds adsorbed to the particles, what is the mass of these compounds ejected each year into the atmosphere?

13. What is the major worry about the increase in particulates in the atmosphere?

Chapter 12


QUANTITATIVE MODELING OF NATURAL CHEMICAL CYCLES


## Introduction

Geochemical models ('global dispersion models', 'element cycling models') describing the cycles of elements at the earth's surface are the most frequent approach to study of the dispersion of materials, natural or pollutant, on a global basis. These models can be used to estimate the importance of modification of natural chemical cycles owing to perturbations introduced by man's activities. Geochemical models generally describe transport paths and fluxes among a limited number of physically well-defined portions of the earth, referred to as spheres or reservoirs. The majority of models are concerned with the description of transfer of material from the continents to the atmosphere or oceans and vice versa. In some cases, sediments are considered a reservoir separate from the oceans. Main transport paths for major elements are rather well known, and natural fluxes among the three reservoirs have been estimated successfully. The quality of these models is controlled by the accuracy of the mass balance for each reservoir. If a steady state is assumed for the model, which is usually the case for natural element cycles on a geologic time scale, there is no accumulation or removal in the reservoir and the mass balance is equal to zero.

In the case of substances in low concentrations, like the heavy metals and synthetic organics, distributions in all the compartments necessary for the model are commonly less well known than for major species. Data for fundamental chemical and biological characteristics of these substances are generally insufficient to predict their behavior in each reservoir. Also, the input of some minor elements to the natural system by man is comparable to the natural input, thus invalidating an assumption of steady state. Global dispersion models of minor elements are, at present, limited and their quality poor, even in the case of the three major reservoirs model.

To describe or predict the consequences of man's input of pollutants into the environment, a three reservoir model is insufficient; therefore, it is necessary to subdivide the system into more reservoirs such as stratosphere, troposphere, oceanic mixed layer, oceanic deep layer, biosphere, sediments, etc.

149

Subdivision greatly complicates description of the system because it is necessary to predict the concentrations of pollutants within compartments and
rates of transfer among compartments, such as mercury concentrations in water,
plankton, and sediments of the ocean, and mercury transfers between these
compartments.

In developing a global model it is necessary to separate the system of
interest from its natural surroundings. The boundaries of a natural system
are defined by the scale of the phenomena of interest and by previous knowledge of possible interactions between the system and its surroundings. Global dispersion models involve consideration of phenomena on a worldwide scale,
resulting in subdivision of the earth into a number of physically well defined
reservoirs or boxes. Thus, the term 'box model' is commonly applied to models
of this type.

The number of reservoirs considered depends on the way in which the substance of interest is distributed throughout the earth's surface environment;
that is, on the number and direction of transport paths for the substance. A
transport path is a directional property of the system, the route by which a
substance is transported from one box to another.

In a simple three-box model involving the global reservoirs of land, atmosphere, and ocean, the major transport agents of materials from one reservoir
to another are streams, groundwater, rain, ice and wind. If, however, the system of interest is subdivided into more boxes, such as subdivision of the atmosphere into stratosphere and troposphere and subdivision of the ocean into
mixed layer and deep layer, then currents and biological processes become
important agents of material transfer. For example, upwelling currents and
settling of biological debris in the ocean are important means of material
transfer between the oceanic mixed and deep layers.

The manner in which a substance is transported from one box to another
depends on its physical-chemical characteristics. A substance with a high
water solubility would be expected to move about the earth's surface via
water transport. On the other hand, a substance with a high solubility in
fatty tissue, such as DDT, would be expected to be involved in biological processes and to be transported by organisms.

Figure 45 demonstrates the distinction between reservoir and compartment
for a three-box, land-atmosphere-ocean global model of an hypothetical sub-

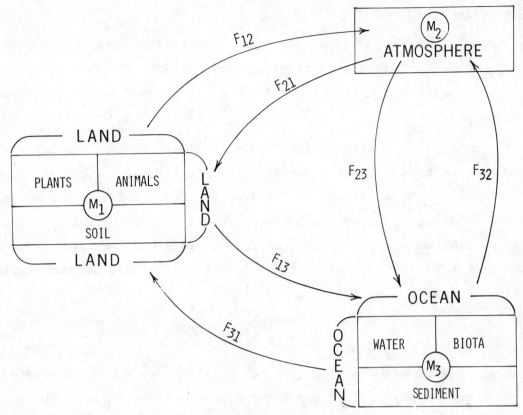

FIGURE 45.  SCHEMATIC DIAGRAM OF A THREE-BOX GLOBAL CYCLING MODEL.
$F_{IJ}$ REPRESENTS FLUX OF A SUBSTANCE BETWEEN RESERVOIRS,
WHERE $I,J = 1,2,3$ AND $I \neq J$.  $M_I$ IS MASS OF SUBSTANCE IN
RESERVOIR, WHERE $I = 1,2,3$.

stance .

Three reservoirs of a substance are shown with inter-reservoir transport
paths.  The ocean reservoir is subdivided into compartments of water, biota,
and sediment;  the land reservoir into compartments of plants, animals and
soil.  Masses of the substance in these compartments are unknown, and although
inter-compartmental transport paths are suspected between some compartments,
no material flux values are known, except for those between reservoirs.  The
flux of a substance is the rate at which the mass of the substance is transferred along a transport path between reservoirs and usually is defined in terms
of mass per unit time (e.g., grams/year$^{-1}$).  Material transport for the sub-

stance in the system of Figure 45 involves (1) gaseous transport from the sea and land surfaces and return in rainfall, (2) river transport of the dissolved and suspended load, and (3) return to the land via uplift of sedimentary rocks.

## Mathematical Considerations

Box models involve simple mathematical formulations, but require the critical judgment of an experienced practitioner. The basic relations used express the mass balance for each box and for all the boxes together (global mass balance).

Interactions of the system with the external environment provide the inputs and outputs to the system. Physical, chemical or biological transformations of the substances in the system lead to production or consumption of substances. The mass balance for a system requires that the rate of change of mass of any substance in the system, $\Delta M_s / \Delta t$, is equal to the input flux $(F_i)$ plus the production term (P) minus the output flux $(F_o)$ and the consumption term (C). The general equation is:

$$\frac{\Delta M_s}{\Delta t} = F_i + P - F_o - C. \tag{1}$$

Usually the terms of the mass balance equation are expressed as mean concentrations of the substance in the various phases considered. Each of the terms in equation (1) is generally composite because there are different possible transport agents controlling the inputs and the outputs in the model, and various processes of removal or production of the substance in the system. Also, the substance may occur in different phases (dissolved chemical substance, suspended particles, living organisms) and as various chemical species (molecules, ions, etc.).

An initial simplification is to consider the total concentration of the substance in a number of limited but representative phases. In the case of mercury, for instance, we need not consider each of the numerous mercury compounds, but simply the total mercury concentration in each phase. For example, in the ocean these phases may be the dissolved material, suspended minerals, and plankton. For each of these phases in a system a mass balance equation can be written, plus a mass balance equation for the whole system. If various chemical species of mercury are considered, a mass balance equation can be written for each of the species in the various phases considered.

Input and output terms are selected for the known transport agents between reservoirs. For processes like river discharge, sedimentation, and upwelling, these terms may be expressed as the product of the total flux of the transport agent times the mean concentration of the substance in that flux. For other processes like exchange between the lower and upper layers of the atmosphere or the ocean, or between sediment pore waters and overlying seawater, terms describing fluxes are more complex.

Production and consumption terms describe how a substance's concentration varies with time within a reservoir. For conserved chemical species, those that are not consumed or produced, the terms P and C are equal to zero and equation (1) contains only the input and output fluxes $F_i$ and $F_o$, respectively.

In a global model with m species and n phases, the number of equations that need to be solved simultaneously is equal to m x n. This number is reduced to n when only the total concentration of a substance is taken into account. Description of the system can be considerably simplified if steady state is assumed. Steady state models are usually satisfactory if interest is in description of the _average_ properties of a system over long periods of time. Most models of world-wide material dispersal are steady state, based on the fact that, from geological evidence, the composition of the atmosphere, ocean and earth's crust have remained constant for at least a few hundreds of millions of years.

Equation (1) is then reduced to:

$$F_i + P - F_o - C = 0, \qquad (2)$$

and for a conservative chemical species:

$$F_i = F_o . \qquad (3)$$

Attempts have also been made to construct time dependent concentration models; for example, of $CO_2$ in the atmosphere, DDT in the surface environment, and of phosphorus in the surface environment. In these models, fluxes for transfer of materials between reservoirs are considered as a function of time (e.g., phosphorus discussion in this chapter).

With this brief introduction to modeling of global systems, we can now consider some mathematical aspects of steady state models.

## Steady State Models

Figure 46 illustrates an example of a mathematical formulation of a three-

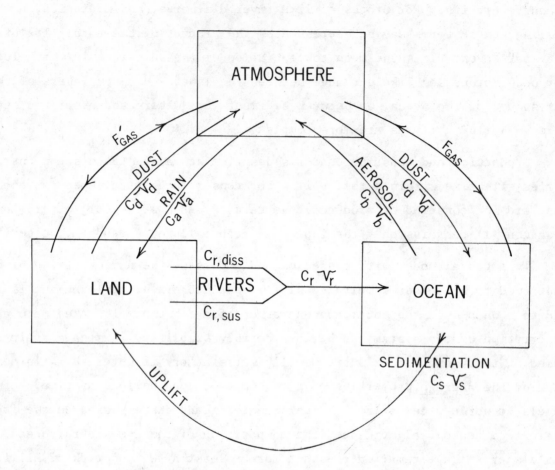

FIGURE 46.  SCHEMATIC DIAGRAM OF A THREE-BOX, STEADY STATE,
GLOBAL CYCLING MODEL OF A SUBSTANCE.

reservoir model (land-ocean-atmosphere) for a conservative substance (no net production or consumption) having a gaseous state.  Carbon dioxide is an example.  The fluxes considered are:

1.  River discharge to oceans:  total flux of the substance equals total concentration in rivers, $C_r$, times total discharge, $\mathcal{V}_r$.  $C_{r,\,diss}$ and $C_{r,sus}$ are, respectively, concentrations of the substance in the dissolved and suspended load of rivers.

2.  Sedimentation in the oceans:  total flux is rate of oceanic sedimentation, $\mathcal{V}_s$, times concentration, $C_s$, of substance in marine sediments.

3.  Continental dust:  total flux of substance to the ocean is flux of dust from continents to ocean, $\mathcal{V}_d$, times concentration, $C_d$, of substance in dust.

4.  Rainfall over continents:  total flux of substance in rain is total rainfall over continents, $\mathcal{V}_a$, times concentration, $C_a$, of substance in rain.

5.  Sea aerosol:  total flux of substance to atmosphere from ocean is total aerosol flux, $\mathcal{V}_b$, times concentration, $C_b$, of substance in aerosol produced by bursting bubbles at sea surface.

6.  Gaseous exchange:  difference in fluxes between atmosphere and land or atmosphere and sea surface and return, $F'_{gas}$ and $F_{gas}$, respectively.

The mass balance equations for the individual boxes and for the whole system, assuming steady state, and assuming that there is net gaseous transport from ocean to land via the atmosphere, are:

1.  Land:  $F'_{gas} + C_a \mathcal{V}_a + C_s \mathcal{V}_s = C_r \mathcal{V}_r + C_d \mathcal{V}_d$;

2.  Ocean:  $C_r \mathcal{V}_r + C_d \mathcal{V}_d = C_s \mathcal{V}_s + C_b \mathcal{V}_b + F_{gas}$;

3.  Atmosphere:  $F_{gas} + C_b \mathcal{V}_b + C_d \mathcal{V}_d = C_a \mathcal{V}_a + F'_{gas} + C_d \mathcal{V}_d$.

Dust flux terms usually cancel out because dust does not undergo chemical modification in the atmosphere.

4.  Whole system -- steady state condition that there is no net flux in or out of the global system.  This steady state model contains seven unknown fluxes ($\mathcal{V}_r, \mathcal{V}_s, \mathcal{V}_d, \mathcal{V}_b, \mathcal{V}_a, F_{gas}, F'_{gas}$) and five unknown concentrations ($C_d, C_a, C_s, C_b, C_r$).  There are, however, two independent relations, thus the model is solvable if only two variables are unknown.

It is possible to simplify these equations by imposing conditions.  For example, for a minor element with low volatility, the terms $F_{gas}$ and $F'_{gas}$ drop out.  If aerosol transport of the element is also negligible, the array of equations reduces to:

1.  Land:  $C_s \mathcal{V}_s = C_r \mathcal{V}_r + C_d \mathcal{V}_d$,

2.  Ocean:  $C_d \mathcal{V}_d + C_r \mathcal{V}_r = C_s \mathcal{V}_s$,

and

3.  Atmosphere:  $C_d \mathcal{V}_d = C_d \mathcal{V}_d$.

In this case, one of the relations is trivial, and only one equation can be used if values for all the variables are unknown.

The following section applies the principles developed to some element cycles.

155

## Examples of Modeling

### Mercury

In Chapter 10, the pre-man and present global cycles of mercury were presented. In this section the steps leading to their construction are outlined.

1. Four major reservoirs and masses were considered:

Atmosphere: Hg mass computed from mass of atmosphere ($5.2 \times 10^{21}$g), density of atmosphere ($1.3 \times 10^3$g/m$^3$), and mean atmospheric mercury content (1 ng/m$^3$). $5.2 \times 10^{21} \div 1.3 \times 10^3 \times 1 \times 10^{-9} = 40 \times 10^8$g.

Oceans: Hg mass computed from total oceanic volume ($1.37 \times 10^{21}$ liters) and mean concentration of dissolved Hg in seawater (0.03 µg/liter). $1.37 \times 10^{21} \times 30 \times 10^{-9} = 411,000 \times 10^8$g. Mercury in suspended matter in the ocean is about $2,100 \times 10^8$g; this value added to the dissolved mass gives a total oceanic Hg burden of about $415,000 \times 10^8$g.

Sediments: Hg mass computed from average values of Hg in shales (150 ppb), sandstones (75 ppb), and carbonates 75 ppb); relative proportions of shale, carbonate, and sandstone in sediments (75:14:11, respectively); and total sedimentary mass ($25,000 \times 10^{20}$g). $(0.75 \times 150 \times 10^{-9} + 0.14 \times 75 \times 10^{-9} + 0.11 \times 75 \times 10^{-9})(25,000 \times 10^{20}) = 3,300,000,000 \times 10^8$g.

Land: Hg mass computed from land area less that covered by ice ($133 \times 10^{16}$ cm$^2$), average worldwide soil thickness (60 cm), soil density (2.5g/cm$^3$), and mean Hg content of soil (50 ppb). $133 \times 10^{16} \times 60 \times 2.5 \times 50 \times 10^{-9} = 100,000 \times 10^8$g.

2. Pre-man cycle: For each reservoir, assuming the pre-man cycle of mercury was in steady-state, we may write the following equations representing fluxes into and out of a reservoir.

$$x_1 + x_3 = x_2 + x_4 \text{ (atmosphere)}$$
$$x_4 + x_7 = x_3 + x_5 \text{ (oceans)}$$
$$x_5 = x_6 \text{ (sediments)}$$
$$x_2 + x_6 = x_1 + x_7 \text{ (land)},$$

where x represents the material fluxes in units of g/yr shown in Figure 37. Thus there are 7 unknowns but only 4 equations. The following three conditions enable the solution of the array of simultaneous linear algebraic equations:

$x_2 = x_4/0.5$; Hg fallout on land and sea surfaces is proportional to their surfaces.

$x_5 = x_6 = 13 \times 10^8$ g/yr; steady-state assumption computed by using the pre-man denudation rate of the continents ($100 \times 10^{14}$ g/yr) and the average Hg content of sediments (132 ppb). ($100 \times 10^{14} \times 132 \times 10^{-9} = 13 \times 10^8$ g/yr). Major assumption is that the rate of supply of 'new' land surface via uplift is equal to rate of denudation of continents.

$x_1 + x_3 = 250 \times 10^8$ g/yr $= x_2 + x_4$; steady-state assumption computed by using the average Hg content of Greenland ice from 800 B.C. to 1952 (0.06 ppb) and assuming that worldwide precipitation was $4.2 \times 10^{20}$ g (cm$^3$/yr), so total mercury flux to earth's surface is $60 \times 10^{-12}$ g Hg/g precipitation x 4.2 $\times 10^{20}$ g precipitation $= 250 \times 10^8$ g/yr

The solution for the pre-man cycle is given in Figure 37.

3. Present cycle: The present cycle of mercury is not in steady state because the land and ocean reservoirs are gaining Hg, whereas the sedimentary rock reservoir is being depleted of mercury. The atmosphere is a steady-state reservoir. To derive the present cycle of Hg, the following calculations and conditions were satisfied:

(a) Flux from oceans to sediments computed on basis of

(1) mass of siliceous and calcareous skeletons deposited annually ($7.5 \times 10^{14}$ g/yr and $1.2 \times 10^{15}$ g/yr, respectively) and average skeletal Hg concentration (0.4 ppm). ($1.95 \times 10^{15} \times 0.4 \times 10^{-6} = 7.8 \times 10^8$ g/yr);

(2) steady-state assumption that river suspended load, calculated from total suspended load of world's rivers ($183 \times 10^{14}$ g/yr) and average Hg concentration of shales (150 ppb) ($183 \times 10^{14} \times 150 \times 10^{-9} = 27 \times 10^8$ g/yr), passes through the ocean without reaction;

(3) steady-state assumption that airborne dust falling on ocean surfact, calculated from continental dust flux ($5 \times 10^{14}$ g/yr), average Hg concentration of shales (150 ppb), and fallout of 70 percent of the dust on the sea surface ($5 \times 10^{14} \times 150 \times 10^{-9} \times 0.70 = 0.53 \times 10^8$ g/yr), passes through the ocean without reaction.

Summation gives: $7.8 \times 10^8 + 27 \times 10^8 + 0.53 \times 10^8 = 35 \times 10^8$ g/yr.

(b) Flux from sediments to land computed on basis of present denudation rate ($240 \times 10^{14}$ g/yr) and average Hg content of sediments (132 ppb). $240 \times 10^{14} \times 132 \times 10^{-9} = 32 \times 10^{8}$ g/yr.

(c) Flux from land to oceans computed by summation of dissolved Hg load of world's rivers, calculated from total yearly runoff ($0.32 \times 10^{20}$ g/yr) and average Hg concentration of world's rivers (0.07 ppb) ($0.32 \times 10^{20} \times 0.07 \times 10^{-9} = 22 \times 10^{8}$ g/yr), and particulate Hg load of world's rivers, calculated from total suspended load ($183 \times 10^{14}$ g/yr) and average concentration of Hg in shales (150 ppb) ($183 \times 10^{14} \times 150 \times 10^{-9} = 27 \times 10^{8}$ g/yr). Summation gives $22 \times 10^{8} + 27 \times 10^{8} = 50 \times 10^{8}$ g/yr.

(d) Mining flux of $90 \times 10^{8}$ g/yr Hg

(e) Total emission flux of $102 \times 10^{8}$ g/yr obtained from summation of atmospheric Hg emission owing to man's activities:

| source | emission rate ($10^8$ g/yr) |
|---|---|
| chlor-alkali production | 30.0 |
| coal-lignite combustion | 27.4 |
| oil-gas combustion | 22.9 |
| cement manufacturing | 1.2 |
| sulfide ore roasting | 20.0 |
| | 101.5 = 102 |

(f) Assumption that the ocean to atmosphere flux increased by 1/3 of the pre-man flux to $223 \times 10^{8}$ g Hg/yr. Using this condition and the above fluxes, the remaining fluxes of Figure 38 can be obtained and the changes in reservoir masses calculated. Justification for this condition is that the present total flux of Hg to the atmosphere from the sea and land surfaces is about $410 \times 10^{8}$ g/yr, a value in accord with an increase of Hg in glacial ice of about two times pre-man concentrations. Weiss, et al. (op. cit.) showed that precipitation on the Greenland ice sheet prior to 1952 contained $0.06 \pm 0.02$ ppb Hg, whereas those waters deposited as ice from 1952-1965 contained $0.13 \pm 0.05$ ppb Hg, an observation in good agreement with the model prediction.

## Phosphorus

In the previous section there was little manipulation of the mercury cycle. Conclusions were drawn chiefly from examination of current and past reservoir sizes and fluxes. However, if there is enough good information (which is rare!), more sophisticated mathematical models can be constructed for use in predicting the consequences of various possible changes in fluxes of a given substance. The potential of such models for important predictions is great, but their use is in its infancy, and is chiefly retarded by lack of basic information. The following discussion attempts to show the basic principles involved in developing such models, and uses the global cycle of phosphorus as an example. Perhaps the chief message of the discussion is to show the areas of current ignorance of information required for accurate prediction.

Figure 47 shows a recent estimate of reservoirs and fluxes of phosphorus (Lerman, Mackenzie and Garrels[*]). (The figure and the following discussion are derived from their work.) Table 23 gives the sources for the sizes of the reservoirs and fluxes and serves admirably to illustrate the ranges of values that have been obtained by various 'authorities'. Note that a Land Reservoir has been established, based on the P content of the soil zone, in turn based on an average thickness of 60 cm for soil of the land areas of the earth. Note also an arbitrary choice of 300 meters for the depth of the Surface Ocean.

The cycle as shown in 'balanced'; that is, the fluxes in and out of each reservoir are equal. The dashed line for 'Mineable P' flux is not included in the balanced cycle, but will be introduced in discussing man's perturbations. Thus the cycle as shown is representative of the supposed long-term steady-state situation of the geologic past.

On the upper left, the flux from soil (Reservoir #2) to plants (Reservoir #3) is estimated on the basis of photosynthetic rates and P content of terrestrial plants. The return flux is supposed equal, with the same rate of release of P by decay and oxidation of organic matter as that caused by photosynthesis.

The total P transferred to the soil from the sediment reservoir is 20 million metric tons per year; 18.3 of this is shown cycling directly back into sediments. This is the P that is fixed in the soil as insoluble Al and Fe phosphates. Its path could have been shown as going into the surface ocean, then to the deep ocean, and then to the sediments, but so far as is known, it does not react en route, and so is shown returning directly to sediments.

Only 1.7 million metric tons of the twenty added to the soil by the uplift and erosion of sedimentary rocks is carried to the Surface Ocean dissolved in streams.

*Lerman, A., Mackenzie, F.T., and Garrels, R.M., 1975. Modeling of Geochemical Cycles: Phosphorus as an Example. in Geol. Soc. Am. Memoir 142, pp. 205-218, E.H.T. Whitten, ed.

159

# GLOBAL CYCLE OF PHOSPHORUS

*Fluxes and reservoir masses in millions of metric tons P (units of $10^{12}$ g).*

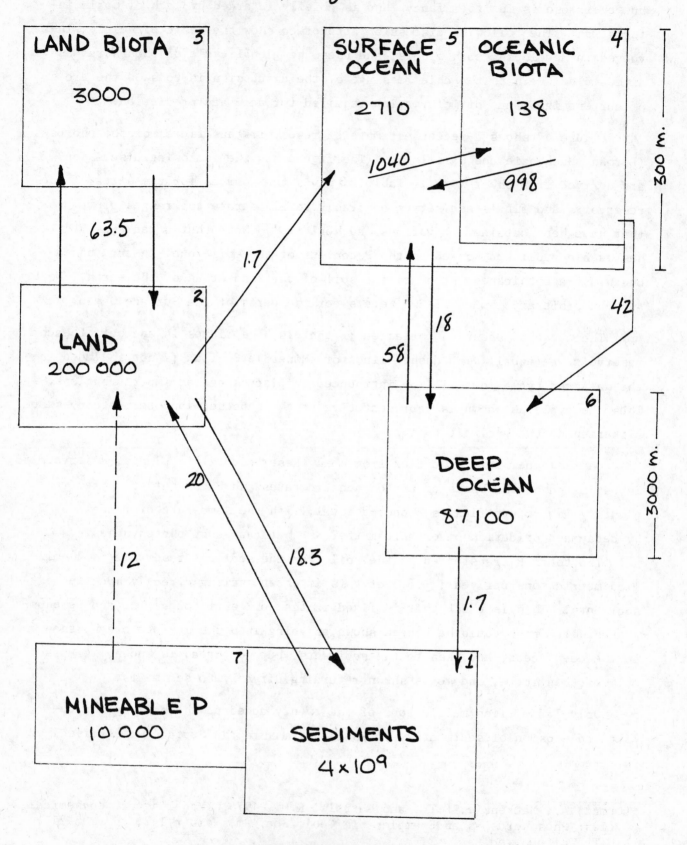

# Figure 47

For the Surface Ocean there are three chief sources of P:  from streams, from upwelling (58 million metric tons), and from decay of marine organisms (998 million metric tons).  A small amount of the surface water P is lost downward by mixing with deeper waters (18 million metric tons);  most of the P moves into the Oceanic Biota via photosynthesis.  The Oceanic Biota lose P when dead organisms sink, decompose, and add soluble P to the Deep Ocean.

Final balance is achieved by depositing 1.7 million metric tons of P in the ocean sediments;  part as P in buried organic materials, part presumably precipitated as solid calcium fluorophosphate $[(Ca_5(PO_4)_3F]$.  This net flux out of the deep ocean neglects a smaller cycle, in which more P than shown is initially buried in the sediments, but some diffuses back into the deep waters as organic material decomposes.  Estimates for this sub-cycle are currently inadequate to permit its inclusion here.

This steady-state system, as indicated, roughly resembles the natural unperturbed system.  Before studying the effect of perturbations on the P system, it is necessary to investigate perturbations of a simplified cycle.

## A Simplified Phosphorus System

For the purpose of illustration, it is possible to consider Reservoirs 4, 5 and 6 of Figure 47 as a closed system.  There is a small flux of dissolved P from streams each year into Reservoir #5, and a comparable loss to sediments $(1.7 \times 10^{12}$ g/yr) from Reservoir #6 coming from Reservoir #4, but the system consisting of the Oceanic Biota, the Surface Ocean, and the Deep Ocean, with its large reservoirs and fluxes, can be studied as if there were no stream input and no loss to sediments, provided the time interval considered is not excessively long.  At any rate, this three reservoir system will be used to illustrate how perturbations of a three reservoir steady-state system can be assessed, operating at the mathematical level of simple linear algebraic relations.

Figure 48 reproduces parts of Figure 47, ignoring stream input and loss to sediments.  As shown, Figure 48A is a steady-state system, with the sum of P fluxes into the reservoirs equal to the sum of P fluxes out of the reservoirs.  Figure 48B sets up terminology for referring to the reservoirs and fluxes.

In cycling studies a usual first approximation is that the flux from a reservoir is proportional to the size of the reservoir.  The rationale for this

161

# A. THE OCEANIC CYCLE OF PHOSPHOROUS

## (Modified slightly to achieve a closed system)

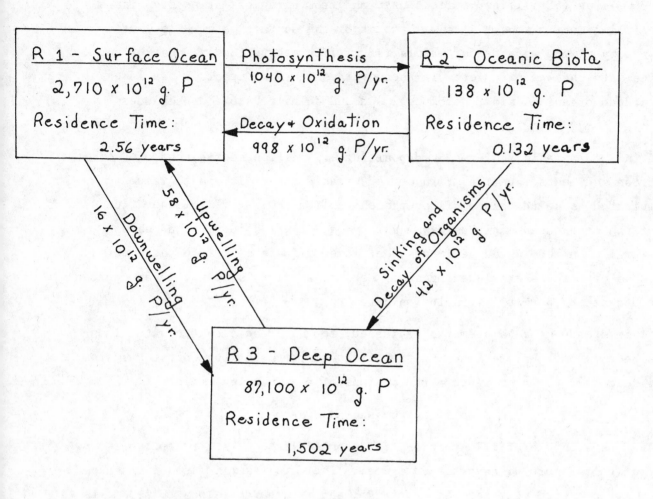

# B. FORMALIZATION OF OCEANIC CYCLE OF PHOSPHOROUS

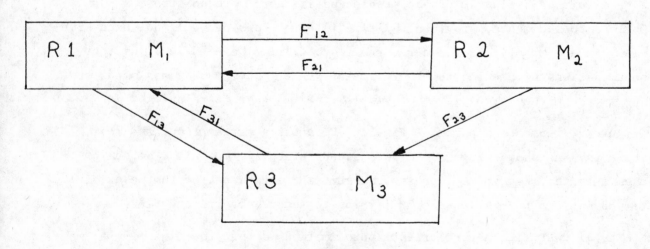

FIGURE 48

assumption can be shown by considering the rate of decay of oceanic biota and release of P ($F_{21}$). Decay is an intensive variable -- the rate per gram of organic material depends on the local conditions, the amount of oxygen present, the bacterial population, etc. Consequently, if there is a very large biomass, and the rate of decay per gram is fixed, the total amount of decay is large. If there is a small biomass, and the rate of decay per gram is the same, the total amount of decay is reduced. If one dead fish decays completely in a month, 10,000 dead fish will also decay in a month, so the decay flux will be 10,000 times greater.

This approximation can be stated formally as, "The flux of a component from a reservoir is proportional to the size of the reservoir". In terms of Figure 48B, $F_{21} \propto M_2$, or $F_{21} = k_{21}M_2$, where k is known as a rate constant.

There are many systems that do not follow this relation, but it is commonly used as a first test of observed relations and as a method for rough prediction.

For the steady-state cycle presented in Figure 48A, the rate constants for the various masses and fluxes can be calculated.

$$k_{12} = F_{12}/M_1 = 1040 \times 10^{12} g\ P/yr/2,710 g\ P = 0.384/yr,$$

$$k_{21} = F_{21}/M_2 = 998 \times 10^{12} g\ P/yr/138 \times 10^{12} g\ P = 7.232/yr,$$

$$k_{13} = F_{13}/M_1 = 16 \times 10^{12} g\ P/yr/2,710 \times 10^{12} g\ P = 0.005904/yr,$$

$$k_{31} = F_{31}/M_3 = 58 \times 10^{12} g\ P/yr/87,100 \times 10^{12} g\ P = 0.000666/yr,$$

$$k_{23} = F_{23}/M_2 = 42 \times 10^{12} g\ P/yr/138 \times 10^{12} g\ P = 0.304/yr.$$

If the sizes of the reservoirs or fluxes are changed for any reason, the assumption is made that the constants do not change.

### Perturbation of the System

Now we are in a position to assess the effects of various types of perturbations of the closed system. The major controls of the present oceanic system are the limiting effects of P and N on the rate of photosynthesis. Let us hypothesize that some other control takes over the photosynthetic rate; perhaps a diminution of light because of emissions of particulate materials, or toxic effects of oil residues in the surface ocean. Further, let us assume that the effect is instantaneous, and that the photosynthetic rate is cut in half overnight, i.e., $k_{12}$ is changed from 0.384/yr to 0.192/yr.

Qualitatively, what should happen? The rate of production of the biomass will be cut in half, but there is no reason to expect that the rate of decay, the rate of sinking of organisms, the rate of upwelling, or the rate of downwelling will be affected. The biomass should shrink; the flux of decay ($F_{21}$) should decrease; the flux of sinking organisms should diminish. Because the total system is closed, i.e., $M_1 + M_2 + M_3$ = a constant, there should be a redistribution of phosphorus among the reservoirs. As the biomass declines in P content, the surface and deep waters should be enriched. It is reasonably evident that a new steady state will be achieved, with lower total P in the biomass, and higher total P in the waters. To solve the problem quantitatively, the relations can be written (see Fig. 48B), using the new value of 0.192 for $k_{12}$:

$$M_1 + M_2 + M_3 = \text{constant (closed system)} = 89948 \times 10^{12} \text{g P} \tag{1}$$

$$F_{12} = 0.192 \, M_1 \tag{2}$$

$$F_{13} = 0.005904 \, M_1 \tag{3}$$

$$F_{23} = 0.0304 \, M_2 \tag{4}$$

$$F_{21} = 7.232 \, M_2 \tag{5}$$

$$F_{31} = 0.000666 \, M_3. \tag{6}$$

For balance between the reservoirs at the new steady state:

$F_{12} = F_{21} + F_{23}$, or from (2), (5) and (4) above,

$0.192 \, M_1 = 7.23 \, M_2 + 0.304 \, M_2$,

$$M_2 = 0.02546 \, M_1. \tag{7}$$

Also, for reservoir balance:

$F_{31} = F_{23} + F_{13}$. From (6), (4) and (3) abo e, $0.000666 \, M_3 =$

$0.3043 \, M_2 + 0.005904 \, M_1$.

Substituting from (7),

$0.000666 \, M_3 = 0.00776 \, M_1 + 0.0059 \, M_1$

$$M_3 = 20.51 \, M_1. \tag{8}$$

Substituting (7) and (8) in (1):

$$M_1 + 0.02546\ M_1 + 20.51\ M_1 = 89,948 \times 10^{12} g,$$

$$M_1 = 4175 \times 10^{12} g,$$

$$M_2 = 106.3 \times 10^{12} g,\qquad\qquad\text{[from (7)]}$$

$$M_3 = 85,667 \times 10^{12} g.\qquad\qquad\text{[from (8)]}$$

The fluxes between the reservoirs are obtained by substituting the values for reservoir masses (M) into equations (2) to (6). The result, shown in Figure 49, is the new steady state condition. Note that the fluxes are not perfectly balanced, a result of the numerical manipulations.

Comparison of Figures 48A and 49 shows that the final effects of an initial, instantaneous lowering of the photosynthetic rate by 50% are to lower the biomass to about 2/3 of its original size and to increase dissolved P in ocean surface waters by 50%. Deep ocean P and the upwelling flux are only slightly changed, but downwelling is increased because of the increase of P in surface water.

Note that, under the assumptions of this model, an initial cut in the rate of photosynthesis by 50% does not halve the size of the biomass, and that the new steady state photosynthetic flux is greater than the value at the time of the rate change. These results are because of feed-back; as the biomass diminishes and dissolved P increases, the photosynthetic _flux_ of P rises, although the rate of photosynthesis _per gram_ of P in surface waters remains at 50% of the value given in Figure 48A.

Nothing has been said about the time required for the achievement of the new steady state, although it should be clear that changes would be rapid immediately after lowering the photosynthetic rate, and that the changes would become slower and slower as the new equilibrium was approached.

### Model with Constant Photosynthetic Flux

A variation of the model system presented would be to decrease instantaneously the photosynthetic flux of P to 50% of its initial value, and then hold the _flux_ constant. As indicated before, the rate constants for the other fluxes should not change. In this case,

$$F_{12} = \text{constant} = F_{21} + F_{23}$$

at the new steady state. Rewriting,

$$F_{12} = k_{21}M_2 + K_{23}M_2.$$

STEADY STATE IN SEAWATER - PHOSPHOROUS SYSTEM

AFTER DECREASE OF PHOTOSYNTHETIC RATE BY 50%

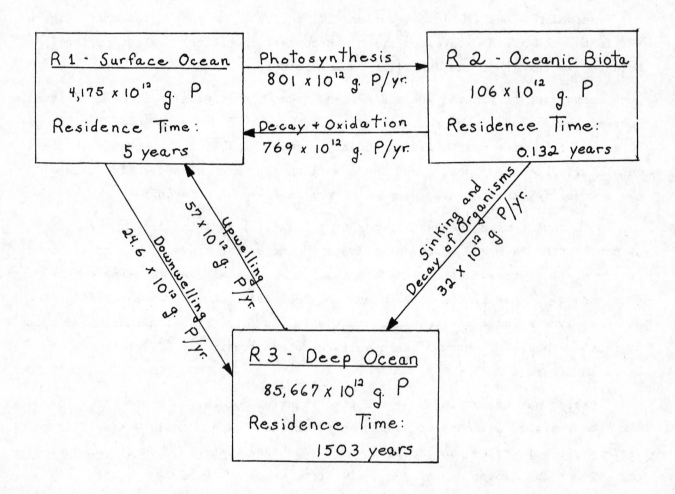

FIGURE 49

The initial photosynthetic flux (Fig. 48A) is $1040 \times 10^{12}$ g P/yr;  the new value is $520 \times 10^{12}$ g P/yr.  Substituting values for $k_{21}$ and $k_{23}$, we obtain

$$520 \times 10^{12} = 7.23 \, M_2 + 0.3043 \, M_2,$$

$$M_2 = 69 \times 10^{12} \text{ g P}.$$

Therefore, changing the photosynthetic flux to half the initial steady state value would halve the biomass.  Values for the remaining masses and fluxes will not be calculated here, but the overall result would be similar to the system shown in Figure 49, but with a smaller biomass and a greater enrichment of P in surface waters.

The relations illustrated by the three reservoir system can be extended to any number of reservoirs and fluxes (see next section), maintaining the principle that flux is proportional to size.  Differential equations can be set up and solved by the computer to give reservoir sizes and fluxes at any time between the perturbation of the system and the new steady state.

## Perturbation of More Complex Models

Lerman, et al. (op cit.), have assessed the effects of several types of perturbations of the P cycle.  The first 'scenarios', to use current governmentese parlance, is to assume that for some reason photosynthesis is suddenly halted on land and in the sea.  This 'Doomsday' situation is equivalent to changing $k_{54}$ and $k_{23}$ of Figure 47 instantaneously to zero.  Figure 50 shows the results.  The land and sea biota continue to decay, but no more biota is produced.  As shown in Figure 50, the oceanic biota would decay within a year, but it would take 200 years for decay of the land biota.  As a result, during 200 years the P content of Ocean Surface water would increase by a factor of 2.5, whereas the P content of the Land and Deep Ocean Reservoirs would be almost unaffected, nor would there be a change in the sediment reservoir.  In the new steady state reached, when terrestrial and oceanic biota are completely decayed, the amounts of P in the Land, Deep Ocean, and Sediment Reservoirs return to the values shown in Figure 47.  However, the amount in the Ocean Surface increases to 8,970 million metric tons, from an initial value of 2,710 million metric tons.  One way of viewing the situation is to see that the decomposition of the oceanic biota would result in a uniform vertical concentration of P in the oceans.

Figure 51 shows predictions of the results of the additions of P to the land as a result of mining of P and adding it to the soil as fertilizer.  An

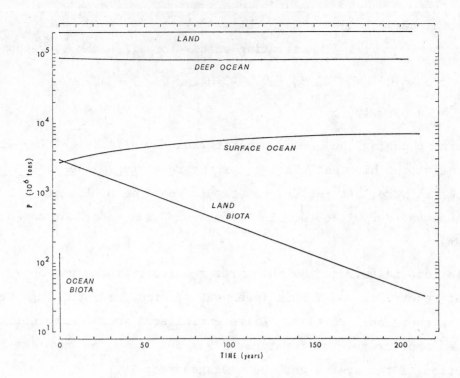

FIGURE 50. DOOMSDAY SCENARIO. P RESERVOIR CHANGES AFTER
ALL ORGANIC PRODUCTIVITY CEASES ON EARTH.

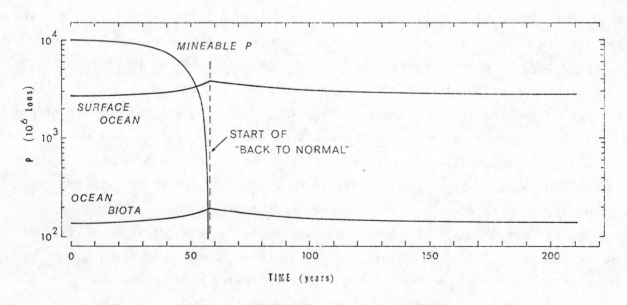

FIGURE 51. P RESERVOIR CHANGES OWING TO AN EXPONENTIALLY
INCREASING RATE OF P MINING AND REMOVAL OF P
BY STREAMS. CONDITIONS BEGIN TO RETURN TO
NORMAL AT 58 YEARS FROM THE PRESENT.

extreme situation is considered -- an exponential increase in the mining and use of P as a land fertilizer, and a total amount of mineable P of 10,000 million metric tons. The Figure shows that the mineable P would be exhausted in about 60 years, and that the effects of its addition to the natural system would be small -- reflected chiefly in an increase in Oceanic Biota and in the concentration of P in Ocean Surface waters. Of course, if nitrogen became limiting, P would simply accumulate in surface waters.

It is hoped that this brief summary of one way of modeling changes in natural systems indicates the techniques that are currently being applied in attempting to predict what will happen in the future. Two major points should be emphasized. First, Table 23 shows the large range of values of fluxes and reservoirs currently estimated by various researchers. Second, mathematical modeling based on the assumption that flux is proportional to reservoir size is a crude first approximation of reality. Information is desperately needed on the actual relations between reservoirs and the fluxes into and from them.

## Table 23

Phosphorus contents of geochemical reservoirs and inter-reservoir fluxes

| Reservoir | P Content (Metric Tons) | | References and Remarks |
|---|---|---|---|
| | Used in this paper | Other estimates | |
| Sediments | $4 \times 10^{15}$ | $7.8 \times 10^{14}$ | Cited in Van Wazer (1961, p. 1285) Poldervaart (1955) |
| | | $8.2 \times 10^{14}$ | Ronov and Korzina (1960) |
| | | $8.37 \times 10^{14}$ | Stumm (1972) |
| Land | $2 \times 10^{11}$ | | Computed from land area (total land less ice, $133 \times 10^6 km^2$), assumed soil thickness (60 cm), density (2.5 g/cm$^3$) and mean P content of crustal material (0.1 wt%, Taylor, 1964); $133 \times 10^6 \times 6 \times 10^{-4} \times 2.5 \times 10^9 \times 1 \times 10^{-3} = 2.0 \times 10^{11}$ tons P. P content of soils reportedly ranges from 0.02 to 0.83 wt% (Fuller, 1972). |

Land biota          $3 \times 10^9$

Computed from nitrogen amount in land biota ($12 \times 10^9$ tons N; Delwiche, 1970) and mean P/N atomic ratio in land plants (1.8/16; Deevey, 1970). $12 \times 10^9 \times 1.8 \times 31/(16 \times 14) = 3.01 \times 10^9$ tons P.

$1.41 \times 10^9$

Computed from carbon amount in land biota ($450 \times 10^9$ tons C; Bolin, 1970) and mean P/C atomic ratio in land plants (1.8/1,480; Deevey, 1970). Computation as in preceding estimate.

$3.43 \times 10^8$

Computed from $CO_2$ amount in living matter on land ($4 \times 10^{11}$ tons $CO_2$; Lemon, 1969) and mean P/C atomic ratio in terrestrial organic matter (1.8/1,480; Deevey, 1970). Computation as above.

$1.95 \times 10^9$   Stumm (1972)

Oceanic biota   $1.38 \times 10^8$

Computed from nitrogen amount in oceanic biota ($1 \times 10^9$ tons, (Vaccaro, 1965) and mean P/N atomic ratio in oceanic biota (1/16) (Redfield and others, 1963); compare Ryther and Dunstan, 1971). $1 \times 10^9 \times 1 \times 31/(16 \times 14) = 1.38 \times 10^8$ tons P.

$1.22 \times 10^8$

Computed from carbon amount in oceanic biota ($5 \times 10^9$ tons C; Bolin, 1970) and mean P/C atomic ratio in oceanic biota (1/106; Redfield and others, 1963). Computation as in preceding estimate.

$1.99 \times 10^8$

Computed from $CO_2$ amount in living ocean biota ($3 \times 10^{10}$ tons $CO_2$; Lemon, 1968) and P/C atomic ratio of 1/106 (Redfield and others, 1963). Computation as above.

$1.24 \times 10^8$   Stumm (1972)

Surface ocean   $2.71 \times 10^9$

Computed from assumed mean concentration of total dissolved phosphorus (25 mg/m$^3$; compare Sverdrup and others, 1942, p. 241) in 300-m-thick water layer, and ocean surface area ($3.61 \times 10^8$ km$^2$). $25 \times 0.3 \times 3.61 \times 10^8 = 2.71 \times 10^9$ tons P.

| | | |
|---|---|---|
| Deep ocean | $8.71 \times 10^{10}$ | Computed from assumed mean concentration of total dissolved phosphorus ($80$ mg/m$^3$; compare Sverdrup and others, 1942, p. 241) in 3,000-m-thick layer, and ocean surface area. Computation as in previous estimate. |
| | $8.1 \times 10^{10}$ | From Stumm's (1972, Fig. 2) presentation of Broecker's (1971) data. |
| Mineable resource | $1 \times 10^{10}$ | Assumed from within the range of the estimates given below. |
| | $3.1 \times 10^{10}$ | Stumm (1972) |
| | $1.8 \times 10^{9}$ | Cited in Ronov and Korzina (1960) |
| | $4.67 \times 10^{10}$ | Van Wazer (1961, 0. 966) |

Fluxes (Metric Tons/Yr)

| | | |
|---|---|---|
| $F_{12}$ | $2 \times 10^{7}$ | Computed from the combined rates of mechanical and chemical denudation of continents (approx. $2 \times 10^{10}$ tons/yr; Garrels and Mackenzie, 1971) and mean P content of crustal material ($0.1$ wt %; Taylor, 1964). This flux includes the dissolved and detrital forms of phosphorus. The fraction of this flux in solution is $1.7 \times 10^{6}$ tons/yr, as explained in the entry $F_{25}$. |
| $F_{23}$ | $6.3 \times 10^{7}$ | Computed from Bolin's (1970) figure for the total amount of carbon fixed annually by land plants (20 to 30 billion tons) and mean P/C atomic ratio in land biota ($1.8 \times 31/(1,480 \times 12) = 6.3 \times 10^{7}$ tons/yr. |
| | $2.29 \times 10^{8}$ | Stumm (1972) |
| | $6.26 \times 10^{7}$ | Computed from rate of photosynthesis on land ($7.3 \times 10^{10}$ tons $CO_2$ yr; Lemon, 1968) and mean P/C ratio (Deevey, 1970) in land biota. Computed as above. |
| $F_{25}$ | $1.7 \times 10^{6}$ | Garrels and others (1973, p. 80). This estimate excludes the agricultural and other man-produced contributions. |
| | $2 \times 10^{6}$ | Stumm and Morgan (1970, p. 550) |

$1.86 \times 10^6$     Stumm (1972). This estimate includes agricultural and other man-produced contributions.

$4.4 \times 10^5$     Cited in Ronov and Korzina (1960)

$F_{32}$    $6.35 \times 10^7$     Estimate balanced by the flux $F_{23}$

$F_{54}$    $1.04 \times 10^9$     Computed from the rate of nitrogen fixation by oceanic biota ($7.5 \times 10^9$ tons N/yr; higher value cited by Vaccaro, 1965, $1 \times 10^{10}$ tons N/yr) and mean P/N atomic ratio in oceanic biota (1/16; Redfield and others, 1963; compare Ryther and Dunstan, 1971). $7.5 \times 10^9 \times 1 \times 31/(16 \times 14) = 1.04 \times 10^9$ tons P/yr.

$9.75 \times 10^8$     Computed from the rate of carbon fixation by oceanic biota ($4 \times 10^{10}$ tons C/yr; Bolin, 1970) and mean P/C atomic ratio in oceanic biota (1/106; Redfield and others, 1963). Computed as above.

$9.61 \times 10^8$     Stumm (1972)

$3.06 \times 10^9$     Computed from the $CO_2$ fixation rate by oceanic biota ($4.6 \times 10^{10}$ tons $CO_2$/yr; Lemon, 1968) and mean P/C atomic ratio, computed as above.

$F_{45}$    $9.98 \times 10^8$     Computed on assumption that 96% of living oceanic biota is being recycled within the upper 300-m-thick water layer.

$F_{46}$    $4.2 \times 10^7$     Differences between the fluxes $F_{54}$ and $F_{45}$.

$F_{56}$    $1.8 \times 10^7$     Computed from mean concentration of total dissolved P in surface ocean (25 mg/m$^3$; see entry "Surface ocean" in this table), water exchange rate between the surface and deep ocean layers (2 m/yr; Broecker, 1971), and surface area of the ocean ($3.61 \times 10^8$ km$^2$). $25 \times 2 \times 10^{-3} \times 3.61 \times 10^8 = 1.8 \times 10^7$ tons P/yr.

$F_{65}$    $5.8 \times 10^7$     Computed as for the flux $F_{56}$, using 80 mg/m$^3$ for phosphorus concentration in deep ocean.

$F_{61}$    $1.7 \times 10^6$     Flux value assumed as balanced by the stream flux to the ocean ($F_{25}$).

$F_{72}$    $1.2 \times 10^7$     Stumm (1972)

$1.5 \times 10^7$     Meadows and others (1972, p. 26; as $PO_4^{3-}$?).

References:

Bolin, B., 1970, The carbon cycle: Sci. Am., v. 223, p. 125-132.

Broecker, W., 1971, A kinetic model for the chemical composition of sea water: Quarter. Res., v. 1, p. 188-207.

Deevey, E. S., Jr., 1970, Mineral cycles: Sci. Am., v. 223, p. 149-158.

Delwiche, C. C., 1970, The nitrogen cycle: Sci. Am., v. 223, p. 137-146.

Fuller, W. H., 1972, Phosphorus element and geochemistry, in Fairbridge, R. W., ed., The encyclopedia of geochemistry and environmental sciences: Van Nostrand Reinhold, New York, p. 942-946.

Garrels, R. M., and F. T. Mackenzie, 1971, Evolution of sedimentary rocks: Norton, New York, 397 pp.

Lemon, E. R., 1968, Carbon fixation and solar energy budget, in Altman, P. L., and D. S. Dittmer, eds., Metabolism (Biological Handbooks): Federation of Am. Soc. for Experimental Biology, Bethesda, Md., p. 487.

Meadows, D. H., D. L. Meadows, J. Randers, and W. W. Behrens III, 1972, The limits to growth: Universe Books, New York, 205 pp.

Poldervaart, A., 1955, Chemistry of the Earth's crust: Geol. Soc. America Spec. Pap., No. 62, p. 119.

Redfield, A. C., B. H. Ketchum, and F. A. Richards, 1963, The influence of organisms on the composition of seawater, in Hill, M. N., ed., The Sea: Interscience, New York, v. 2, p. 26-77.

Ronov, A. B., and G. A. Korzina, 1960, Phosphorus in sedimentary rocks: Geochemistry, No. 8, p. 805-829.

Ryther, J. H., and W. M. Dunstan, 1971, Nitrogen, phosphorus, and eutrophication in the coastal marine environment: Science, v. 171, p. 1008-1013.

Stumm, W., 1972, The acceleration of the hydrogeochemical cycling of phosphorus, in Dyrssen, D., and D. Jagner, eds., The changing chemistry of the oceans: Wiley, New York, p. 329-346.

Sverdrup, H. U., M. W. Johnson, and R. H. Fleming, 1942, The oceans: Prentice-Hall, Englewood Cliffs, N. J., 1087 pp.

Taylor, S. R., 1964, Abundance of chemical elements in the continental crust: a new table: Geochim. Cosmochim. Acta, v. 28, p. 1273-1285.

Vaccaro, R. F., 1965, Inorganic nitrogen in sea water, in Riley, J. P., and G. Skirrow, eds., Chemical oceanography: Academic Press, New York, v. 1, p. 365-408.

Van Wazer, F., ed., 1961, Phosphorus and its compounds: Interscience, New York, v. 2, 1091 pp.

## Lead Metabolism in Humans - A Three Reservoir Model

The uptake and metabolic balance of lead in a healthy adult have been studied using the stable isotopes of Lead $Pb^{204}$ and $Pb^{207}$ as tracers (Rabinowitz, et al.[*]). The results of this work can be interpreted in terms of the three-reservoir model shown in Figure 52. The human body is divided into three compartments: bones, blood and soft tissue. Approximately two-thirds of the lead assimilated comes from food and drink (about 10% of the total ingested per day); the remainder is inhaled. Lead leaves the body by excretion from the blood into urine and in sweat, hair, nails and feces. The bones are the principal repository for lead in the human body; the residence time of lead in bones is 78 years, whereas that of blood and soft tissue is only about 30 days.

Using the model of Figure 52, it is possible to estimate the changes that would occur in lead distribution in the human body if lead intake by inhalation or ingestion from food and drink were suddenly increased. The increased lead could come, for example, from eating food cooked in earthenware with defective lead glaze. For a person whose lead absorption has increased from 0.050 mg/day to one mg/day, what are the distribution and fluxes of lead in the new steady state?

The rate constants of the system shown in Figure 52 are:

$$k_{12} = \frac{20}{1800} = 0.0111 \qquad k_{31} = \frac{7}{200,000} = 3.5 \times 10^{-5}$$

$$k_{21} = \frac{8}{700} = 0.0114 \qquad k_1(\text{urine}) = \frac{38}{1800} = 0.0211$$

$$k_{13} = \frac{7}{1800} = 3.89 \times 10^{-3} \qquad k_2(\text{feces etc.}) = \frac{12}{700} = 0.0171.$$

At the new steady state, the condition is that assimilation of lead is equal to removal, thus

$$F_1'(\text{urine}) + F_2'(\text{feces, etc.}) = 1000 \mu g/day.$$

$$0.0211 M_1' + 0.0171 M_2' = 1000,$$

$$M_1' = 47,393 - 0.810 M_2',$$

[*]Rabinowitz, M.R., Wetherill, G.W., and Kopple, J., 1973. Lead metabolism in normal human: Lead isotope studies. Science 182, 725-727.

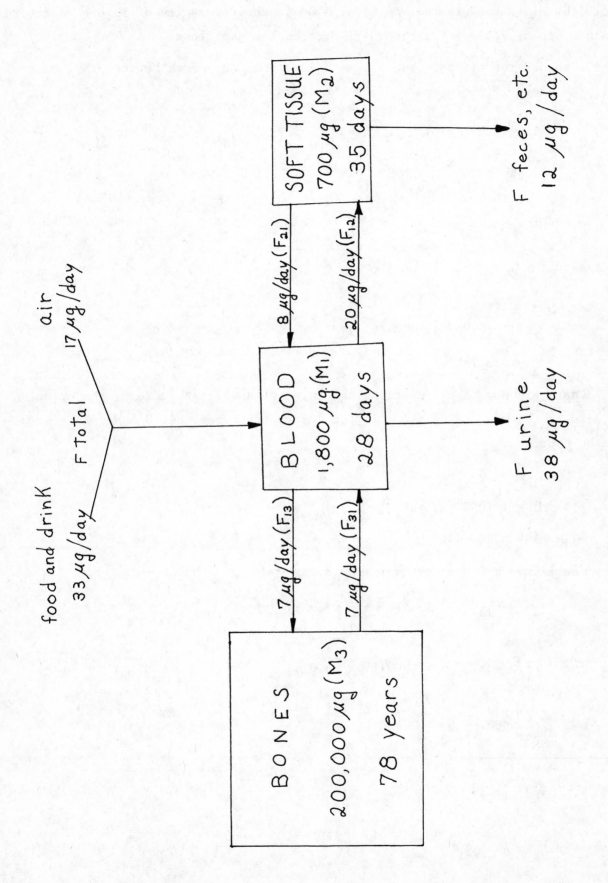

FIGURE 52.   LEAD METABOLISM IN THE NORMAL HUMAN

where the M' values are the masses of lead in the reservoirs of Figure 52 after the new steady state is reached. F' denotes the new flux.

A second condition is that flux of lead back and forth between the bones ($M_3'$) and blood ($M_1'$) of Figure 52 is the same.

$$F_{13}' = F_{31}'; \quad k_{13}M_1' = K_{31}M_3'.$$

$$M_1' = \frac{3.5 \times 10^{-5}}{3.89 \times 10^{-3}} = M_3',$$

$$M_1' = 9 \times 10^{-3} M_3'. \tag{2}$$

Also,

$$F_{12}' - F_{21}' = F'(\text{feces, etc.}),$$

$$0.0111M_1' - 0.0114M_2' = 0.017M_2',$$

$$M_2' = 0.3895M_1'. \tag{3}$$

There are now three equations and three unknowns; therefore, the masses of lead at the new steady state in the reservoirs bones, blood and soft tissue can be obtained:

$$M_1' = 36,040 \ \mu g,$$

$$M_2' = 14,038 \ \mu g, \text{ and}$$

$$M_3' = 4,004,444 \ \mu g.$$

The fluxes between reservoirs are:

$$F_{13}' = k_{13}M_1' = 3.89 \times 10^{-3} \times 36,040$$

$$= 140 \ \mu g/day,$$

therefore $\quad F_{12}' = k_{12}M_1' = 0.0111 \times 36,040$

$$= 400 \ \mu g/day,$$

$$F_{21}' = k_{21}M_2' = 0.0114 \times 14,038$$

$$= 160 \ \mu g/day;$$

$$F'(\text{urine}) = F'(\text{total}) + F_{31}' + F_{21}' - F_{13}' - F_{12}'$$

$$= 1000 + 140 + 160 - 140 - 400$$

$$= 760 \quad \text{g/day},$$

$$F'(\text{feces,etc.}) = F'_{12} = F'_{21} = 400 - 160 = 240.$$

The new steady state is shown in Figure 53.

Notice that the bones, blood and soft tissue contain increased amounts of lead; the concentration of lead in the bone reservoir, as well as in blood and soft tissue, is increased about 20 times, probably enough to result in calcium loss by replacement.

This example provides some idea of how increased lead intake can affect lead storage and metabolism in the human body. It also shows how simple models can predict quantities of trace element uptake in organs and transfer rates of the element through the body. Coupled with toxicological information these models may be useful in predicting physiological effects of increased doses of trace elements on the human body.

The model originally presented, if examined closely, however, cannot be entirely accurate. As shown, there is no accumulation in the bones; the flux in equals the flux out. But the huge bone reservoir (200,000 µg) tells us that lead tends to increase in the bones, even though it may be in a steady state with regard to ingestion, blood, soft tissue and excretion. Consequently, the flux into the bones probably is never balanced by return from the bones; the flux in must be greater than the flux out. This is another way of saying that the flux out of bone is not proportional to their content, as is assumed in this type of modeling. The lead becomes 'fixed', i.e., it is not in a mobile form that would tend to migrate out as the concentration in bone increases.

A better model, probably within the limits of error of the analyses on which Figure 52 is based, would be to have ingestion of lead exceed excretion by 7 µg/day, the flux into the bones, and have a zero or negligible flux from bones back to blood. In this model, lead going into the bones could be treated as if it were excreted.

## Summary

Some examples of quantification of models of element cycles were presented in this chapter to illustrate the types of data used and some of the mathematical formulations involved. Examples of both steady- and transient-state models were presented. The time-dependent models are of most interest because they provide one means of estimating man's future impact on natural chemical cycles.

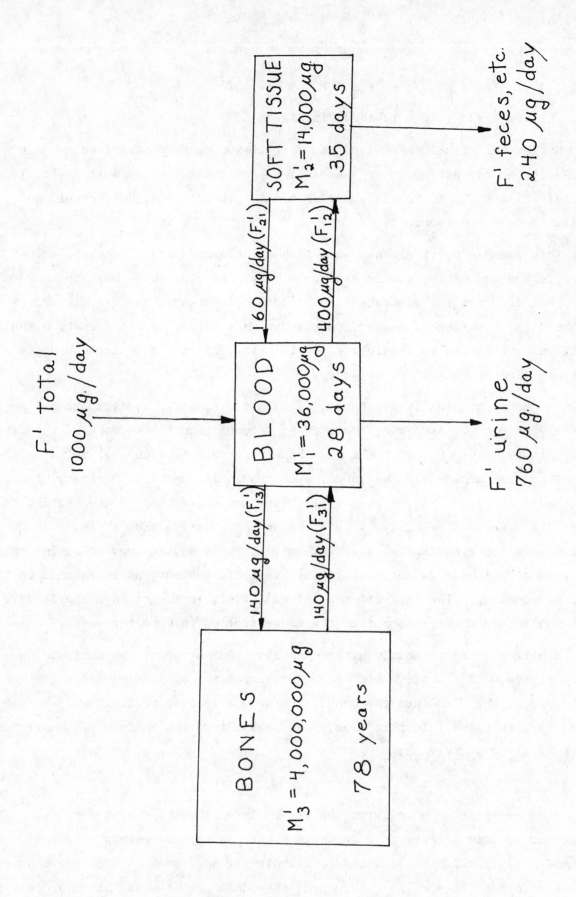

FIGURE 53. LEAD METABOLISM IN HUMAN AFTER 20x INCREASE IN LEAD ASSIMILATED

Study Questions

1. On the basis of the diagram provided for the 'quantitative cycling of P', compare the residence time of P in surface ocean waters with the residence time of the oceanic living biomass.

2. Using the diagram of question #1, calculate $K_{23}$ and $K_{32}$.

3. If the terrestrial photosynthesis rate ($K_{23}$) is decreased instantanteously by 50%, calculate the size of the terrestrial biomass reservoir at the new steady state, ignoring the fluxes into the land reservoir from sediments and out of the land reservoir in streams or as solid Al and Fe phosphates.

4. Choose the P reservoir that would be affected most by a continuing use of fertilizer at an increasing rate.

5. Referring to the cycle of Pb in the human body, estimate the total Hg in micrograms in blood and soft tissue, if the residence time of Hg in the body is 3 months.

6. Do you think the proportion of Hg lost through sweat, hair and nails would be greater or less than the proportion of lead lost this way?

APPENDIX I

ROCKS AND MINERALS

Igneous Rocks are those that have crystallized from the molten state.  They
can be derived either from the melting of pre-existing igneous rocks, from the
melting of sedimentary rocks, or from a mixture of both.

Sedimentary Rocks are those that have been deposited by wind or water or ice,
and can have as their sources either pre-existing igneous or sedimentary rocks.
They almost always occur in layers with two long dimensions and one short one.
The layers were originally deposited nearly flat, but in the course of geologic
time may become folded, broken, and contorted.

Minerals are naturally occurring chemical compounds.  These compounds are
usually found in small grains (less than 1 cm).  Each grain has a uniform chemical composition throughout.  Of the thousands of minerals, we will concentrate
on only about 15.

Rock Composition:  Both igneous and sedimentary rocks average about 50% oxygen,
and most of the minerals can be considered to be made up of oxides of the metals;
i.e., sodium oxide ($Na_2O$), calcium oxide ($CaO$), ferric iron oxide ($Fe_2O_3$), aluminum oxide ($Al_2O_3$), silicon oxide ($SiO_2$).

Solid Solution:  In many minerals, two or more metals may combine with oxygen
or other elements in almost any ratio without changing the structure or general
properties of the mineral.  For instance, minerals can be found that have all
compositions from $CaCO_3$ to $SrCO_3$, in which the ratio of Ca/Sr ranges from infin-
ity to zero.  Such variations in composition without change in structure are
called solid solutions.

Some Important Rocks and Minerals

## A. IGNEOUS ROCKS

| Rock | Chief Minerals | Composition | As Oxides |
|---|---|---|---|
| Granite | quartz | $SiO_2$ | $SiO_2$ |
| | K-feldspar | $KAlSi_3O_8$ | $K_2O \cdot Al_2O_3 \cdot 6SiO_2$ |
| Solid | Na-feldspar | $NaAlSi_3O_8$ | $Na_2O \cdot Al_2O_3 \cdot 6SiO_2$ |
| Solution | Ca-feldspar | $CaAl_2Si_2O_8$ | $CaO \cdot Al_2O_3 \cdot 2SiO_2$ |
| | Mica | $KAl_3Si_3O_{10}(OH)_2$ | $K_2O \cdot 3Al_2O_3 \cdot 6SiO_2 \cdot 2H_2O$ |
| Basalt | Na-feldspar | $NaAlSi_3O_8$ | $Na_2O \cdot Al_2O_3 \cdot 6SiO_2$ |
| Solid | Ca-feldspar | $CaAl_2Si_2O_8$ | $CaO \cdot Al_2O_3 \cdot SiO_2$ |
| Solution | Pyroxene | $FeSiO_3$ | $FeO \cdot SiO_2$ |
| | | $MgSiO_3$ | $MgO \cdot SiO_2$ |

Solid / Solution

## B. SEDIMENTARY ROCKS

| | | | |
|---|---|---|---|
| Sandstone (15%)* | quartz | $SiO_2$ | $SiO_2$ |
| | K-feldspar | $KAlSi_3O_8$ | $K_2O \cdot Al_2O_3 \cdot 6SiO_2$ |
| Carbonate rock (15%)* | Calcite | $CaCO_3$ | $CaO \cdot CO_2$ |
| | Dolomite | $CaMg(CO_3)_2$ | $CaO \cdot MgO \cdot 2CO_2$ |
| Evaporite (5%)* | Gypsum | $CaSO_4 \cdot 2H_2O$ | $CaO \cdot SO_3 \cdot 2H_2O$ |
| | Halite | $NaCl$ | |

Shale (65%)* Clay minerals, i.e. $Me^+_{0.32}[(Fe^{+3}_{0.21}Mg_{0.42}Al_{1.51})$

$$(Si_{3.68}Al_{0.32})]O_{10}(OH)_2$$

*Percentage of total sedimentary rock mass.

APPENDIX II

GLOSSARY AND DEFINITIONS

Absolute humidity = the weight of water vapor in a given volume of air.

Concentration = Basically the fraction of the total of a substance made up by one component; i.e., stream water averages 15 parts per million by weight of calcium (15 ppm Ca). Concentration also commonly given in moles per liter, in weight percent, in parts per thousand by weight (‰), or parts per billion (ppb) by weight.

Density = mass per unit volume; i.e., $g/cm^3$ - one gram of pure water has a volume of 1 $cm^3$, so its density is one.

Emissions = Flux to the environment owing to man's activities; usually refers to flux of particulate and gaseous materials to atmosphere.

Flux = the rate at which a substance is being transferred from one reservoir to another. For example, $CO_2$ is getting transferred at the rate of about $50 \times 10^{14}$ moles/yr from the atmosphere to plants.

Gram Atomic Weight = atomic weight of element in grams; i.e., atomic weight oxygen = 16.1; 1 gram atomic weight = 16.1 grams.

Gram Molecular Weight = molecular weight of compound expressed in grams; i.e., one gram molecular weight of nitrogen ($N_2$) = 28 grams.

Half-life = Time taken for ½ of an original total number of atoms to undergo change: e.g., time for decay of ½ of an original number of atoms of a radioactive element to another product; time to produce a 50-50 mixture of new and old water by addition of new water to old water in a basin at steady state--the percentage of original water diminishes logarithmically, with one-half of the original water removed every half-life.

Ion = an element or compound that has gained or lost an electron, so that it is no longer neutral electrically, but carries a charge. There is essentially no change in mass; calcium for example, occurs in stream water as calcium ions, $Ca^{++}$, but each ion still has an atomic weight of 40.

One Mole = one gram atomic weight of an element or 1 gram molecular weight of a compound.

Pressure = Force/unit area; i.e., 1 atmosphere pressure = 1 $kg/cm^2$.

Relative humidity = ratio (usually in %) of water vapor in air to water vapor in air at saturation. Relative humidity decreases if a parcel of air is heated without addition of water vapor.

Reservoir = total amount (usually a mass or a number of moles) of a substance in the atmosphere, in the oceans, in sedimentary rocks, etc. For instance, there are about $1400 \times 10^{20}$ g of water in the ocean reservoir, and $0.38 \times 10^{20}$ moles of oxygen in the atmospheric reservoir.

Residence time = total mass of a substance in a reservoir divided by the rate of inflow or outflow. For instance, the residence time of water in a 20 gallon bathtub which is being fed 5 gallons/minute, and out of which 5 gallons/minute is draining, is 20 gallons/5 gal/minute = 4 minutes. This means any parcel of water has an average residence time in the tub of four minutes.

Sink = the nature of the storage of the flux of a material into a reservoir. For example, plants are a sink for atmospheric $CO_2$, because they abstract it and fix it in their tissues.

Wastes = Residues produced and not recycled owing to man's agricultural, industrial, etc. activities.

Water saturated air = air containing the maximum amount of $H_2O$ gas it can hold at a given temperature and pressure.

1 $m^3$ air = 1293g;  1ng (nannogram)solid/$m^3$ air = .0007ppb

1 kilogram = 1000 grams = 2.2 pounds

1 manpower = 2500 kcal/day = average person's intake of food

1 kilocalorie = heat to raise 1000g $H_2O$ 1°C.

1 kilowatt-hour = 860 kilocalories

1 kilocalorie = 4167 joules

Some useful conversions, Metric to English

1 meter = 39.34 inches

1 kilometer = 0.6 miles

Average depth of ocean = 4 km = 2.5 miles

Base of stratosphere = 12 km = 7 miles

One meter/second = 3600 meters/hour = 2.2 mph

1/2 kilogram = 1 pint = 1 pound

1 kilogram = 1 quart = 2 pounds

3785 cubic centimeters = 1 gallon

SELECTED BIBLIOGRAPHY

Atmospheric Pollution; Wilfred Bach, McGraw-Hill Book Co., New York, 144 pp.,
    1972.

The Biosphere; Scientific American Book, W.H. Freeman & Co., San Francisco,
    134 pp., 1970.

Carbon monoxide balance in nature; B. Weinstock and H. Niki, Science, 176:
    290-292, 1972.

Carbon monoxide in rainwater; J.W. Sinnerton, R.A. Lamontagne, V.J. Linnen-
    bom, Science, 172: 943-945, 1971.

The Changing Chemistry of the Oceans; D. Dyrssen and D. Jagner, eds., Wiley
    Interscience, New York, 365 pp, 1972.

The Chemical Elements of Life; Earl Frieden, Scientific American, 227 (1);
    52-60, 1972.

Chemical Wastes in the sea: new forms of marine pollution; P.A. Greve,
    Science, 173: 1020-1022, 1971.

Chromium; Series on Medical and Biologic Effects of Environmental Pollutants,
    National Academy of Sciences, Washington; 155 pp, 1974.

The Circulation of the Atmosphere; Edward N. Lorenz, American Scientist, 54:
    402-420, 1966.

Climate and Weather; Hermann Flohn, McGraw-Hill Book Co., New York, 253 pp.,
    1969.

Coal Facts, 1974-1975; National Coal Association, Washington, 95 pp., 1975.

Conservation of Natural Resources; G.H. Smith, ed., 4th ed., John Wiley &
    Sons, New York, 685 pp., 1971.

The Control of Environment; J.D. Foslansky, ed., North-Holland Pub. Co.,
    Amsterdam, (Fleet Academic Editions, New York), 112 pp., 1967.

Dimensions of the Environmental Crisis; J.A. Day, F.F. Fost, P. Rose, eds.,
    John Wiley & Sons, New York, 212 pp., 1971.

The Earth and Human Affairs; Committee on Geological Sciences, Division of
    Earth Sciences, National Research Council - National Academy of Sciences,
    published by Canfield Press, San Francisco, 142 pp., 1972.

The Encyclopedia of Oceanography; Encyclopedia of Earth Sciences Series,
    Vol. 1, Rhodes Fairbridge, ed., Reinhold Pub. Co., New York, 1021 pp.,
    1966.

Energy and the Future; A.L. Hammond, W.D. Metz, and T.H. Maugh II; American
    Association for the Advancement of Science, Washington, 184 pp., 1973.

Energy and Power; Scientific American Book, W.H. Freeman & Co., San Francisco, 1971.

Energy, Economic Growth and the Environment; S.H. Schurr, ed., Resources for the Future Inc., The Johns Hopkins Press, Baltimore, 232 pp., 1972.

The Energy Environment in Which We Live; David M. Gates, American Scientist, 51: 327-348, 1963.

Energy Exchange in the Biosphere; David M. Gates, Harper & Row, New York, 151 pp., 1963.

Energy; Sources, Use and Role in Human Affairs; Carol and John Steinhart, Duxbury Press, No. Scituate, Mass., 362 pp., 1974.

Environment, Resources, Pollution & Society; W.W. Murdoch, ed., Sinauer Associates, Stamford, Conn., 440 pp., 1971.

Environmental Geoscience: Interaction between Natural Systems and Man; A.N. Strahler and A.H. Strahler, Hamilton Pub. Co., Santa Barbara, 511 pp., 1973.

Evolution of Sedimentary Rocks; R.M. Garrels and F.T. Mackenzie, W.W. Norton & Co., New York, 397 pp., 1971.

Extended industrial revolution and climate change; W.R. Frisken, Transactions of the American Geophysical Union (EOS); 52: 500-508, 1971.

Future climates and future environments; F. Kenneth Hare, Bulletin American Meteorological Society, 52: 451-456, 1971.

Geology, Resources and Society: An Introduction to Earth Science; H.W. Menard, W.H. Freeman & Co., San Francisco, 621 pp., 1974.

The Global Circulation of Atmospheric Pollutants; Reginald E. Newell, Scientific American, 224 (1): 32-42, 1971.

Guide to National Petroleum Council Report on United States Energy Outlook; National Petroleum Council, Washington, 40 pp., 1973.

The Heat and Water Budget of the Earth's Surface; David H. Miller, Advances in Geophysics, Vol. 11, Academic Press, New York, pp. 175-302, 1965.

Impingement of Man on the Oceans; Donald W. Hood, ed., Wiley-Interscience, New York, 738 pp., 1971.

Inadvertent Climate Modification, Report of the Study of Man's Impact on Climate; The M.I.T. Press, Cambridge, Mass., 308 pp., 1971.

Introduction to the Atmosphere; Herbert Riehl, McGraw-Hill Book Co., New York, 2nd ed., 516 pp., 1971.

Introduction to Environmental Science and Technology; John Wiley & Sons, New York, 404 pp., 1974.

Lead-Airborne lead in perspective; Committee on Biologic Effects of Atmospheric Pollutants, Division of Medical Sciences, National Research Council, National Academy of Sciences, 1972.

The Limits to Growth; D.L. Meadows, J. Randers, W.W. Behrens III, Potomac Associates Book, Washington, D.C., 205 pp., 1972.

To Live on Earth: Man and His Environment in Perspective; Sterling Brubaker, Resources for the Future, The Johns Hopkins Press, Baltimore, 202 pp., 1972.

Trace Elements in the Atmosphere; H. Israël and G.W. Israël, Ann Arbor Science Pub. Inc., Ann Arbor, Michigan, 158 pp., 1974.

Trace Elements in the Environment; E.L. Kothny ed., American Chemical Society, Washington, Advances in Chemistry Series 123, 149 pp., 1973.

Man and the Ecosphere; Readings from Scientific American, W.H. Freeman & Co., San Francisco, 307 pp., 1971.

Man and His Geologic Environment; D.N. Cargo and B.F. Mallory, Addison-Wesley, Reading, Mass., 548 pp., 1974.

Man and His Physical Environment; G.D. Mackenzie and R.O. Utgard, Burgess Pub. Co., 1972.

Mankind at the Turning Point; M. Mesarovic and E. Pestil; E.P. Dutton & Co., New York, 210 pp., 1974.

Man-made climatic changes; Helmut E. Landsberg, Science, 170: 1265-1274, 1970.

Man's Impact on Environment; T.R. Detwyler, ed., McGraw-Hill Book Co., New York, 731 pp., 1971.

Man's Impact on Terrestrial and Oceanic Ecosystems; W.H. Matthews, F.E. Smith, E.D. Goldberg, eds., M.I.T. Press, Cambridge, Mass. 540 pp., 1971.

Man's Inadvertent Modification of Weather and Climate; Council on Environmental Quality, Bull. Am. Meterological Society, 51: 1043-1047,

Manganese; Series on Medical and Biologic Effects of Environmental Pollutants, National Academy of Sciences, Washington, 191 pp., 1973.

Materials and Man's Needs; Committee on The Survey of Materials Science and Engineering, National Academy of Sciences, Washington, D.C., 1974.

Minerals Yearbook 1972; Vol. 1, Metals, Minerals and Fuels. U.S. Dept. of the Interior, U.S. Government Printing Office, Washington, 1370 pp.

Models of Doom: A Critique of the Limits to Growth; H.S.D. Cole, C. Freeman, M. Jahoda and K.L.R. Pavitt, Universe Books, New York, 244 pp., 1973.

Non-Ferrous Metals: A Survey of Their Production and Potential in The Developing Countries; U.N. Industrial Development Organization, New York, 188 pp., 1972.

The Ocean;   Scientific American Book, W.H. Freeman & Co., San Francisco, 259 pp., 1969.

Oceanography: An Introduction to the Marine Environment;   Peter K. Weyl, John Wiley & Sons, New York, 535 pp., 1970.

Oceanography;   Readings from Scientific American, W.H. Freeman & Co., San Francisco, 417 pp., 1971.

Oceans;   K.K. Turekian, Prentice Hall, New York, 120 pp., 1968.

Orientations in Geochemistry;   Panel on Orientations in Geochemistry, U.S. Nat'l. Committee for Geochemistry, Nat'l. Research Council., National Academy of Sciences, Washington, D.C., pp. 53-160., 1973.

Our Energy Supply and its Future;   Battelle Research Outlook, vol. 4 #1, 40 pp., 1972.

Our Geological Environment;   J.S. Watkins, M.L. Bottino, M. Morisawa, W.B. Saunders Co., Philadelphia, 519 pp., 1975.

Petroleum in the Marine Environment;   National Academy of Sciences, Washington, 107 pp., 1975.

Physical Climatology;   William D. Sellers, Univ. of Chicago Press, Chicago, 272 pp., 1962.

Principles for Evaluating Chemicals in the Environment;   National Academy of Sciences, Washington, 454 pp., 1975.

Readings in the Earth Sciences;   Vol. 1 and 2, Sci. Am. Resource Library, W. H. Freeman & Co., San Francisco, 1969.

Statistical Abstract of the U.S.;   U.S. Bureau of the Census, Washington, D.C., 1014 pp., 1973.

Soil, sink for atmospheric CO;   R. Richard and H. Gilluly, Sci. News, 101: 286-287, 1972.

The Surface of the Earth;   Arthur L. Bloom, Prentice Hall, Englewood Cliffs, N.J., 152 pp., 1971.

Understanding Environmental Pollution;   M.A. Strobbe, ed., C.V. Mosby, St. Louis, 357 pp., 1969.

Understanding the Earth: A Reader in the Earth Sciences;   I.G. Glass, P.J. Smith, R.C.L. Wilson, eds., The M.I.T. Press, Cambridge, Mass., 355 pp., 1971.

Vanadium;   Series on Medical and Biologic Effects of Environmental Pollutants, National Academy of Sciences, Washington, 117 pp., 1974.

Water, the Web of Life;   C.A. Hunt and R.M. Garrels, W.W. Norton & Co., New York, 308 pp., 1972.

Answers to Study Questions

Chapter 1

1. Coal, oil, gas.

2. Sulfur gases and particulates ($SO_2$ gas and $H_2SO_4$ droplets), nitrogen oxides (NO and $NO_2$), carbon oxides (CO and $CO_2$), "trace metals" (Pb, Zn, Cd, etc.), particulates (tiny particles of fused siliceous glass, "fly ash").

3. Because piston gasoline engines develop such high temperatures that $N_2$ in the air is oxidized to NO and $NO_2$. Diesel engines have low nitrogen oxide emissions.

4. Because (a) the relative importance of energy sources 25 years hence cannot be predicted, and the type of waste product differs with each, and (b) the degree of control of emissions depends on future technology and environmental concerns.

5. Because current policy is to store and "manage" the wastes by putting them into steel tanks whose tops are exposed at the earth's surface. As the amount of such wastes increases, the possibilities of accidents causing their release inevitably increases. Also, much of the waste is highly radioactive, and the current storage policy would require careful and complicated guardianship for centuries.

6. $^{235}U$, the fuel for conventional reactors, is in relatively short supply, with exhaustion within a few generations foreseeable. The breeder reactor would utilize $^{238}U$, which is more than 100 times as abundant as $^{235}U$. (Waste disposal problems similar or worse.)

7. The first three, if completely developed, would not even supply current world energy demands. Solar energy has the potential of supplying energy requirements for the foreseeable future, but the collecting areas required are large; an area about equal to half the state of Nevada would have to be covered by collectors to supply total present U.S. energy demand. Also, there are many problems yet to be solved in storing solar energy to yield constant supply.

8. All prophets say turn of the century is earliest possible time. Technical breakthroughs are required. However, amount of energy potentially available is enormous. Problems of radioactive waste depend on processes developed, but usually predicted to be fewer than with current fuels.

9. (a) $2.33 \times 10^{13}$ KWH x 860kcal/KWH = $2 \times 10^{16}$ kcal.

   (b) Per capita total consumption of energy = $\dfrac{2 \times 10^{16} \text{kcal/yr}}{210 \times 10^6 \text{ people}}$ =

$95.2 \times 10^6$ kcal/person/yr.  Energy per capita for food per year:

$$2500 \text{ kcal/day} \times 365 \text{ days} = 9.125 \times 10^5 \text{ kcal/person/yr.}$$

Total energy consumption per capita per year/Food energy consumption per capita per year:

$$\frac{95.2 \times 10^6 \text{ kcal/person/yr (total)}}{9.125 \times 10^5 \text{ kcal/person/yr (food)}} = 104 \text{ times as much energy used per person as required from food.}$$

10.  $\dfrac{2 \times 10^{16} \text{ kcal/yr}}{0.30} = 6.7 \times 10^{16}$ kcal/yr world.

$6.7 \times 10^{16}$ kcal/yr/$3.5 \times 10^9$ people $= 19.1 \times 10^6$ kcal/person/yr.

11.  $\dfrac{19.1 \times 10^6 \text{ kcal/person/yr}}{2000 \text{ kcal/person/day} \times 365 \text{ days/yr}} = 26.2$.

12.  $\dfrac{104}{26.2} \times 6.7 \times 10^{16}$ kcal/yr (world) $= 266 \times 10^{15}$ kcal/yr, or

$266 \times 10^{15}/67 \times 10^{15} = 4$ times present world level.

13.  The percentage of metals in mineral deposits is declining, and required depth of mining is increasing.  Result:  more energy required per pound of metal extracted.

As water pollution is being fought, more water treatment plants are being constructed, costing energy for their materials and operations.

Broadly speaking, almost all environmental control systems cost energy for construction and operation (i.e., emission controls on cars, removal of gases and solids from smoke, reclamation of strip-mined land).

14.  Homes and businesses consume only 30% of total energy production, so home consumption is not more than perhaps 20%.  A relatively small proportion of homes are in areas where significant wind or solar energy can be developed.  So a figure of perhaps 10% or less of total energy might be developed by home devices.

15.  $\dfrac{210 \text{ kg/individual/yr}}{600 \text{ kg/individual/yr}} \times 100 = 35\%.$

Emissions enter the atmosphere;  junk (cars, appliances, etc.) goes to dumps;  some discarded materials may be used in landfill projects;  slag and tailings commonly are dumped.

$$\frac{7.5 \text{ kg/individual/yr} \times 10^3 \text{ g/kg} \times 200 \times 10^6 \text{ people}}{200 \times 10^6 \text{ metric tons} \times 10^6 \text{ g/t}} \times 100 = 0.75\%.$$

Note:  This value does not represent total contribution of the iron and steel industry to emissions;  materials other than iron are used in the manufacturing of steel--some wastes from these materials are vented to the atmosphere.

Chapter 3

1. $1031g/cm^2$ x area of earth's surface = $1031g/cm^2$ x $5.1$ x $10^{18}cm^2$ = $5.26$ x $10^{20}g$.

2. $5.26$ x $10^{20}g/29g/mole$ = $1.81$ x $10^{20}moles$.

3. (a) $1.81$ x $10^{20}moles$ x $0.7808$ = $1.41$ x $10^{20}moles$

   (b) At. wt. N = 14, $N_2$ = 28. $1.41$ x $10^{20}$ x $28$ = $39$ x $10^{20}g$.

4. At a given temperature a given volume of air contains a fixed number of moles of gas. If Y is the volume in $cm^3$ containing one mole, then 1 $cm^3$ contains 1/Y moles. Density is weight/$cm^3$. Therefore, for dry air, the density is
$$1/Ymoles/cm^3 \times 29g/mole = 29/Yg/cm^3.$$
For pure water vapor, the density is
$$1/Ymoles/cm^3 \times 18g/mole = 18/Yg/cm^3.$$
For a mixture of 96% dry air and 4% water vapor, the density is
$0.96$ x $29/Yg/cm^3$ + $0.04$ x $18/Yg/cm^3$ = $27.84/Y$ + $0.72/Y$ = $28.56/Yg/cm^3$.
Therefore the ratio of the density of moist air to dry air is $28.56/Yg/cm^3$/ $29/Yg/cm^3$ = $0.98$.

5. New Volume = Old Volume x $\frac{(273 + 35)}{(273 + 25)}$ = $\frac{308}{298}$ = $1.03$ x old volume or density decrease $\cong$ 3%.

6. Trade Winds are cool and dry. As they move toward equator they warm and pick up water vapor. Thus a given volume of air may decrease in density by 4 or 5%. It therefore tends to rise until it reaches an elevation where its density is equal to that of surrounding air.

7. During the flight of the missile (roughly 1 hour from pole to equator), the rotation of the earth will have moved the submarine 1,000 miles to the east (the earth rotates west to east). Thus the missile will land approximately 1,000 miles west of the submarine, or roughly 15° of longitude.

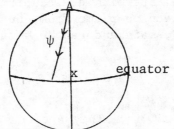

A = missile site

X = submarine

$\psi$ = missile path

This problem, of course, illustrates the tendency of winds moving toward the equator to veer to the west, and those moving from the equator to veer to the east -- the effect of the Coriolis force.

8. The troposphere is heated from the ground, so that low elevation air ordinarily is heated, becomes less dense, rises and mixes. In the stratosphere, temperature increases upwards, so that the air tends to remain where it is, and mixes only slowly. The basic cause is the decreasing upward temperature

of the troposphere, and the increasing upward temperature of the stratosphere.

9. Because most solar energy is not absorbed by the atmosphere, and reaches
the earth's surface as visible light. There it is absorbed, and partially
reheated as longer heat waves. These heat waves are absorbed by water
vapor and especially by $CO_2$. So the lower atmosphere is heated by conduc-
tion from the ground and by absorption of heat by $CO_2$ and water vapor.

10. Because of absorption of ultraviolet light by oxygen to make ozone releasing heat. This reaction is an important shield of the earth's surface from ultraviolet light from the sun.

11. As indicated, warmer air is usually found closest to the ground, heated
by ground radiation. Such air rises and mixes. In an inversion a mass
of cold air accumulates at the land surface, which is overlain in the
troposphere by warm air. This creates a stable situation until the cold
air mass moves out horizontally or is eventually heated enough to rise
and mix. The cold air mass accumulates emissions. Low areas adjacent to
mountains are characteristically susceptible to inversions, for cold air
tends to slide down the mountain slopes at night in quite weather.

12. West flying planes should fly S and then W; East flying planes should
fly N and then E. These routes (see Fig.13) should give the greatest wind
assistance at jet flight heights of about 12 km.

13. $52 \times 10^{20} \times 0.002 = 0.104 \times 10^{20}$ g.

14. $0.104 \times 10^{20}/4.46 \times 10^{20} = 0.023$ years or about 1.2 weeks or 8 days.

15. Radius of earth at 45°N = 2,800 miles.
$(x^2 + x^2 = 4000^2; \quad x = 2800$ miles).

    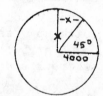

    $2\pi \times 2800 = 18,000$ miles or 30,000 km circumference at
    latitude 45°N.

    Average wind velocity (Fig. 13) = 70 km/hr. 30,000 km/70 km/hr =
    428 hrs. = 18 days.

16. No; should be rained out first, but with dry conditions could be distributed continent-wide in a widening plume from source. N.B. From
time to time, almost all of U.S. mainland can be smoggy.


Chapter 4

1. $Cl^-$, $Na^+$, $SO_4^=$, $Mg^{++}$, $Ca^{++}$, $K^+$, $HCO_3^-$. They are conservative because their
ratios are constant, showing that their residence times are longer than
the times for complete mixing.

2. They must be much more saline. To stay where they are they must be denser
than cold (2 °C) deep water; because density decreases with temperature
and increases with salinity, they must have at least enough excess salinity

to compensate for the 58° temperature difference (double normal salinity or more).

3. Nitrogen, phosphorus and silica.

4. Because oxygen is consumed in oxidizing falling organic material to carbon dioxide with release of nutrients. The process is essentially complete at 800 meters.

5. By upwelling of deep waters and by additions from rivers. Upwelling tends to be localized, as off Peru, where winds blow surface waters offshore, causing their replacement by deeper nutirent-rich waters. Consequently, the supply of nutrients is very unevenly distributed, and areas far from upwelling or river mouths are virtual 'deserts'.

6. Low. They tend to be in areas where evaporation exceeds precipitation, causing increased salinity and sinking of surface water. The Sargasso Sea in the Atlantic, E of Bermuda, is a prime example. Silica content of surface waters is very low -- about 0.01 ppm.

7. The equatorial currents are driven by the westerly component of the NE and SE trades. The Gulf Stream and Kuroshio move east in the belt of westerlies.

8. Only in the thin lighted zone of the oceans can photosynthesis take place. The photosynthetic microorganisms are the primary food of the food chain. Also, deep water pollution may require hundreds of years to work its way to the surface.

9. Reasonable prediction is that it would be higher, because most of the year Hawaii is in the trade wind belt, where evaporation exceeds precipitation. In fact, because of currents, it is about average.

10. There would be no upwelling to restore nutrients, and the surface layers would eventually become entirely depleted. However, there are important contributions of N, P and Si from the land via rivers, so the supply would not be entirely cut off.

11. (a) Total N in upwelling: 200 cm water ($\cong$ 200 g) rises/sq cm ocean surface/yr; $200 \text{ g H}_2\text{O} \times \dfrac{20 \times 10^{-6}\text{g N}}{1000\text{g H}_2\text{O}} \times 3.6 \times 10^{18}\text{cm}^2 =$ $1.44 \times 10^{13}$ g N upwelled/yr.

(b) Total N delivered by streams: $1 \text{ ppm NO}_3 = 1 \times \dfrac{14 \text{ (at wt N)}}{62 \text{ (mol wt NO}_3)} =$ 0.226 ppm N in streamwater. $0.226 \times 10^{-6}\text{g/g} \times .35 \times 10^{20}\text{g H}_2\text{O/yr} =$ $7.9 \times 10^{12}$ g N/yr from streams.

(c)  Ratio:  $1.44 \times 10^{13}$ g N/$7.9 \times 10^{12}$ g N = 1.8 = upwelling/streams.

12.  Inject the poison into the deep, dense water flowing westward out of the Mediterranean.  This water sinks in the Atlantic water until it forms a deep layer.  Thus the poison would be isolated from the surface for hundreds of years.

13.  Ocean average depth/rate of fall:

$4 \times 10^{5}$ cm/500 cm/yr = 800 years.  Note especially the implication that materials can be deposited far from the place they entered the ocean under the influence of ocean currents.  Also, this settling rate is probably close to the minimum.  Many skeletons of organisms settle 100 times faster.

14.  Yes, just as in the atmopshere.  Water from the N Equatorial Current tends to mix northward;  from the S Equatorial Current southward.  Currents moving S in the N hemisphere along eastern margins of the oceans are deflected W along the equator.

15.  No.  For example, deep N Atlantic water extends far S of the equator; Antarctic deep water far N of the equator.

16.  The western side.  A bottle on the eastern side might well move off to the right of the current and get into the Sargasso Sea gyre, where it would circle indefinitely.

17.  Average maximum wind velocities in the troposphere are about 120 km/hr; for ocean surface currents like the Gulf Stream, 6 or 8 km/hr, a ratio of about 20:1.  The average ratio between atmosphere and ocean probably is 100/1, or more.  Also, lateral dispersal is much greater in atmosphere than in oceans because of much more rapid mixing of air masses.

18.  The oceanic system is commonly predictable for periods of years;  the atmospheric system for days (with a wide range for each, depending on local conditions).

19.  A low tidal range, to minimize the amount of oxygenated seawater available beneath the fresh water layer, a shallow depth at the mouth of the estuary or fjord, much greater depths within the estuary or fjord, organic-rich fresh water entering the estuary or fjord.


Chapter 5

1.  About 1/3.  The U.S. is typical.  It averages about 70 cm (27 inches) of rain per year, and just about 23 cm runs into the oceans.

2.  About 100 cm, or $100 \times 3.6 \times 10^{18}$, or $3.6 \times 10^{20}$ g from the whole ocean.

3.  It costs about 9.72 kcal to evaporate 1 mole of water (18 g). So the total energy required is $3.6 \times 10^{20}$ g/18g/mol x 9.74 kcal/mol = $1.95 \times 10^{20}$ kilocalories.

4.  $1.95 \times 10^{20}$ kcal/yr/365 days/yr = $5.3 \times 10^{17}$ kcal/day; $5.3 \times 10^{17}$ kcal/day/$3.5 \times 10^{9}$ persons = $15.1 \times 10^{7}$ kcal/day/person.

5.  $15.1 \times 10^{7}$ kcal/day/person/2500 kcal/person/day = 60000 times the daily requirement.

6.  Yes. Breaking waves, or whitecaps, eject seawater droplets into the air. The water of the droplets evaporates, leaving minute particles of seasalt. These tiny particles are wafted up into the atmosphere. Most return to the sea, but the remainder carry significant amounts of seasalts onto the land.

7.  They are very like, but not identical to, the dissolved salts in the sea. That is to say that they are dominantly NaCl, with subordinate Ca and Mg sulfates.

8.  Most of the material added to the oceans comes from the dissolved and suspended materials brought down by streams (about 80%). The Antarctic and Greenland ice caps contribute significant amounts of ground-up rock, released from these sources when the icebergs melt. Underground waters that reach the sea directly may contribute about 5% as much dissolved material as do streams. Erosion of the shore by waves also adds material, but it is a small fraction of that added by streams. Other sources are wind-blown dust, materials from the interior of the earth (most obviously from volcanoes), and a tiny amount of cosmic dust (with an occasional large meteorite). There is also, of course, continuous interchange of gases between the atmosphere and seawater.

9.  Because at low flow, between rains, the streams are fed by waters that trickle through the soil into the streams. Such waters have had opportunity to react with the soil minerals and dissolve elements from them.

10. They abstract $CO_2$, according to the reaction

$$H_2O + Ca(or~Mg)CO_3 + CO_2 = Ca^{++}_{dissolved} + 2HCO_3^{-}~_{dissolved}.$$

Note that the dissolved bicarbonate is 50% derived from the atmosphere and half derived from the carbonate minerals.

11. Silicate minerals occur as simple metal silicates (i.e., $MgSiO_3$), or more importantly, as metal-aluminum-silicates (i.e., Na-feldspar, $NaAlSi_3O_8$). Simple silicates weather according to reactions of the type:

$$MeSiO_3 + 2CO_2 + H_2O = Me^{++}_{dissolved} + 2HCO_3^{-}~_{dissolved} + SiO_2~diss.$$

Thus they go entirely into solution. Note that <u>all</u> the $CO_2$ required to dissolve them comes from the atmosphere. Aluminosilicates react (using

Na-feldspar, a typical mineral) to yield dissolved metal, bicarbonate, and silica, and a solid residue, typified by the clay mineral
kaolinite.

$$2NaAlSi_3O_8 + 2CO_2 + 3H_2O = \underset{kaolinite}{Al_2Si_2O_5(OH)_4}\text{ solid} + 2Na^+_{\text{dissolved}} +$$

$$2HCO_3^-\text{ dissolved} + 4SiO_2\text{ dissolved}.$$

Note that <u>all</u> the $CO_2$ used comes from the atmosphere.

12. By reversal of the weathering reactions in the oceans or within the
oceanic sediments after they are deposited.

13. The most important is photosynthesis, which uses carbon dioxide and
releases oxygen. $CO_2 + H_2O = CH_2O + O_2$. 99% of the organic material
synthesized is destroyed by bacterial decay and oxidation. Other important reactions are oxidation of ferrous silicates and of ferrous sulfide
(pyrite).

$$2FeSiO_3 + 1/2O_2 = Fe_2O_3\text{ solid} + 2SiO_2\text{ dissolved}$$

$$4FeS_2 + 15O_2 + 8H_2O = 2Fe_2O_3\text{ solid} + 8H_2SO_4\text{ dissolved}.$$

14. When the $H_2SO_4$ is formed in the soil, it quickly reacts with soil minerals
to release cations, i.e.:

$$CaCO_3 + H_2SO_4 = Ca^{++}_{\text{dissolved}} + SO_4^=\text{ dissolved} + CO_2\text{ dissolved} + H_2O$$

The $CO_2$ in turn can react:

$$H_2O + CaCO_3 + CO_2 = Ca^{++}_{\text{dissolved}} + 2HCO_3^-\text{ dissolved}.$$

Silicates may also be attacked by $H_2SO_4$ to yield metal ions, sulfate
ions, and dissolved silica.

15. Total $CO_2$/amount removed/yr = $\dfrac{0.00054 \times 10^{20}\text{ moles C}}{0.025 \times 10^{14}\text{ moles C/yr}}$ = 22,000 years.

16. Presumably not, because it is known that photosynthesis has been occurring continuously for 3 billion years or so. A feed-back mechanism must
operate at about the same rate as the $CO_2$ removal reaction.

17. Probably the $CO_2$ is restored to the atmosphere by the oxidation of
fossil organic material during rock weathering. The percentage of
fossil organic material in old rocks being weathered is just about
equal to that in new rocks being deposited. Thus the feed-back would
be of just the right magnitude to maintain present $CO_2$ levels.

18. $CO_2$ would be depleted in the atmosphere; the rate of photosynthesis
would decline so that less organic material would be produced, thus
reducing the new organic residue remaining for deposition.

19. Yes, there are many possibilities. If atmospheric $CO_2$ were reduced, the ocean, currently approximately in equilibrium with the atmosphere, would begin losing $CO_2$ back to the atmosphere. Because the oceans contain 60 times as much $CO_2$ as the atmosphere (chiefly as dissolved bicarbonate), it would act to 'buffer' the removal of $CO_2$ from the atmosphere, and extend the depletion time manyfold. Also, lowering of atmosphere $CO_2$ pressure would tend to convert carbonate minerals in rocks to silicate minerals, releasing $CO_2$.

20. To form the excess gypsum it was presumably necessary to convert a substantial fraction of the sulfur from pyrite in rocks to sulfate, i.e.,

$$4FeS_2 + 15O_2 + 8H_2O = 2Fe_2O_3 + 8H_2SO_4 \text{ (using oxygen)}$$

The sulfuric acid then could react with $CaCO_3$ to form gypsum.

$$H_2O + H_2SO_4 + CaCO_3 = CaSO_4 \cdot 2H_2O + CO_2.$$

The released $CO_2$ could then be photosynthesized to organic material and buried in sediments, restoring $O_2$ to the atmosphere.

$$CO_2 + H_2O = CH_2O + O_2.$$

21. The weathering of rocks, transportation of materials to the oceans, deposition of sediments, their uplift and re-erosion have transferred amounts of material through the atmosphere and ocean many times greater than their present contents of these materials. If a steady state had not persisted, both atmosphere and oceans would have achieved compositions untenable for living creatures.

22. At the moment, knowledge of the feed-back mechanisms of nature is commonly not sufficient to give definitive answers. However, by comparing natural flux rates of materials to the current or projected ones, some idea of the degree of man's interference can be gained. Also, as well be demonstrated later, some of man's emissions, for example to the atmosphere, have not increased atmospheric content of these gases measureably. In some cases the rate of addition is so high that the observed constancy of the atmosphere demonstrates that removal mechanisms are keeping pace with additions.

23. (a) Water-logged soils are anoxic: the species would be chiefly reduced forms of the compounds $CH_4$, $CO$, $NH_3$, $H_2S$. $CO_2$ occurs as well, although it is a relatively oxidized form of C.

    (b) $CO_2$, $NO_3^-$, $SO_4^=$.

Chapter 6

1. $5000 \times 10^{12}$ moles/yr $\times 110$ kcal/mol $= 5.5 \times 10^{17}$ kilocalories.

2. Total radiation received by earth $= 1.3 \times 10^{21}$ kcal/yr. 47% reaches the surface, or $6.1 \times 10^{20}$ kcal. So, $100 \times 5.5 \times 10^{17}/2.4 \times 10^{21} = 0.09\%$.

3. All but the amount represented by the burial of organic materials in sediments as a residue of decay and oxidation, or

$$2.5 \times 10^{12} \text{ moles/yr} \times 110 \text{ kcal/mol} = 2.75 \times 10^{14} \text{ kilocalories} =$$

$$2.75 \times 10^{14}/5.5 \times 10^{17} = 5 \times 10^{-4} = 0.05\%.$$

4. $4.4 \times 10^{-2} \text{ kcal/yr} \times 5.1 \times 10^{18} \text{ cm}^2 = 2.24 \times 10^{17} \text{ kcal/yr}.$

5. No

6. Greatest production of methane on land is from waterlogged soils, swamps, peat bogs, and especially from rice paddies. In such soils there is no oxygen, and methane is produced by bacterial decomposition of organic material to methane and carbon dioxide: $2CH_2O = CH_4 + CO_2$. The methane is released to the atmosphere because the anaerobic environment in the soil is close to the atmosphere-soil interface. In the oceans, wind and waves keep the water well aerated, so initial methane formed by decomposition of organisms does not reach the atmosphere.

7. Much slower (about 50 times). Methane has a residence time of 3.6 years in the atmosphere, whereas carbon monoxide, with a larger total input, is renewed every 5 weeks or so.

8. They are estimated largely on emissions from waterlogged soils; small fluxes from 'normal' soils may have been missed, and could add up to large totals because of the tremendous areas involved.

9. $3.75 \times 10^{16}$ moles C/$2500 \times 10^{12}$ moles $CO_2$/yr (terrestrial photosynthesis) = 15 years.

10. Total C fixed = $2500 \times 10^{12}$ moles/yr; methane flux to atmosphere $145 \times 10^{12}$ moles/yr; CO flux $20 \times 10^{12}$ moles/yr, so

$$(145 \times 10^{12} + 20 \times 10^{12})100/2500 \times 10^{12} = 7\%.$$

11. $280 \times 10^{12}$ moles/yr $\times 100/3.75 \times 10^{16}$ moles living biomass = 0.75%.

12. No; estimates of living biomass differ by more than 20%; mass is changing also from deforestation, changes in types and acreages of crops, etc.

13. Total moles = moles/g seawater x ocean area x thickness of layer =

$$2.3 \times 10^{-6} \text{ moles} \times 3.6 \times 10^{18} \text{ cm}^2 \times 10,000 \text{ cm} = 8.23 \times 10^{16} \text{ moles (assume}$$

$1 \text{ cm}^3$ seawater = 1 g).

14. No. $280 \times 10^{12}$ moles/yr $\times 100/8.3 \times 10^{16} = 0.34\%$/yr. This would correspond to a change in $HCO_3^-$ from 140 to 140.5 ppm. In ten years the increase in the mixed layer might be detectable (from 140 to 145 ppm ($HCO_3^-$), if there were no mixing with deeper waters.

15. Not very likely. Total $CO_2$ entering and leaving oceans each year in a balanced system is about 8300 x $10^{12}$ moles. So 280 x $10^{12}$ moles/yr is 280 x $10^{12}$ x 100/8300 x $10^{12}$ = 3.4%. But there aren't nearly enough sampling stations to estimate total flux in or out to within 3%.

16. Size of the marine biomass apparently is limited by availability of N and P, and can't increase its size, whereas terrestrial biomass presumably is not so closely controlled by nutrients.

17. One way of illustrating is as follows: The present ratio of oceanic $CO_2$ (chiefly $HCO_3^-$) to atmospheric $CO_2$ (as $CO_2$) is

$$\frac{320 \times 10^{16}}{5.4 \times 10^{16}} = 60.$$ If $CO_2$ in the atmosphere were suddenly doubled,

the ratio would fall to: $\frac{320 \times 10^{16}}{10.8 \times 10^{16}} = 30.$ Eventually, the ratio would

return to 60, so $\frac{(320 \times 10^{16} + x)}{(10.8 \times 10^{16} - x)} = 60$, where x is the number of moles

lost or gained to reachieve equilibrium. Solving the equation gives $x = 5.8 \times 10^{16}$ moles $CO_2$ (or $HCO_3^-$). $CO_2$ remaining in the atmosphere would

be $10.8 \times 10^{16}$ moles - $5.38 \times 10^{16}$ moles = $5.42 \times 10^6$ moles.

In other words, atmospheric $CO_2$ would return to within much less than 1% of what it was before. However, the time scale for complete return is probably of the order of 1,000 years, the mixing time of oceanic deep waters. Note that oceanic content would increase from
$320 \times 10^{16}$ to $325 \times 10^{16}$ moles = about a 1% change.

18. If photosynthetic rate were increased in $CO_2$, let us say to 3000 x $10^{12}$ moles/yr from the present estimate of 2500 x $10^{12}$ moles/yr, oxidation and decay might not keep pace. If so, the amount of organic matter remaining as a residue would increase, and its burial would store $CO_2$. Certainly this mechanism is occurring in terrestrial lakes and in estuaries today, to an important but unknown extent, as organic material from eutrophication accumulates. Also could accumulate in living vegetation with long lifetimes.

19. Do it yourself!

20. (a) 0.058 x $10^{16}$ moles living biomass/0.25 x $10^{16}$ moles decay/yr = 0.23 years for decay of living biomass.

    (b) Only 0.058 x $10^{16}$ moles would be added to the atmosphere's 5.4 x $10^{16}$ moles, a trivial increase.

    (c) Effect on atmospheric $O_2$ would be nil.

    (d) N and P in the mixed layer would be increased by decay of organisms.

21. $35 \times 10^{16}$ moles$/0.5 \times 10^{16}$ moles/yr = 70 years for complete decay.

22. $35 \times 10^{16}$ moles, an increase to $(35 + 5.4) = 40.4 \times 10^{16}$ moles.

23. $$\frac{(320 \times 10^{16} \text{ moles in ocean now} + x \text{ from atm.})}{(40.4 \times 10^{16} \text{ moles in atm.} - x \text{ to ocean})} = 60$$

$320 \times 10^{16} + x = 2424 \times 10^{16} - 60x$

$61x = 2104 \times 10^{16}$

$x = 34.49 \times 10^{16}$ moles.

So $40.4 \times 10^{16} - 34.5 \times 10^{16} = 5.9 \times 10^{16}$ moles left in atmosphere; ocean would rise to $354 \times 10^{16}$ moles. The ppm change in the atmosphere would be from the present 320 ppm to

$$320 \times \frac{5.9 \times 10^{16}}{5.4 \times 10^{16}} = 350 \text{ ppm.}$$

Note importance of rates in assessing actual effects. If $CO_2$ from biomass were all released to atmosphere in 70 years, and 1000 years would be required for re-equilibration with the ocean, the level of $CO_2$ in the atmosphere would rise nearly to the value of $40.4 \times 10^{16}$ moles indicated above. This would be a content of about

$$320 \times \frac{40 \times 10^{16}}{5.4 \times 10^{16}} \cong 2400 \text{ ppm,}$$

which would certainly have important consequences.

## Chapter 7

1. No, they all have small residence times, so, as shown before, evidence of their accumulation in atmosphere would be apparent by this time.

2. Probably yes. There are already several instances of $CO_2$ accumulations in industrial areas during atmospheric inversions that have resulted in many deaths, directly and indirectly. Furthermore, an initial exposure to high levels seems to predispose an individual to serious effects as a result of a later exposure.

3. Because, in the U.S., there is current emphasis on coal as a source of energy, because of the oil shortage, and most coals are high-class emitters of sulfur gases.

4. Probably yes. Several processes seem to be on the verge of being economically feasible, especially because of the rising price of oil.

5.  It has been found that they oxidize sulfur to $H_2SO_4$ droplets, which are
    extremely toxic.  Also, the control devices increase fuel consumption,
    a fact that was of much public concern during the 1974 gasoline shortage.

6.  Because part of the rained-out sulfate comes from sea-salt aerosols,
    and can be considered as $CaSO_4$ particles.  Also, $H_2SO_4$ reacts with $NH_3$
    (or $NH_4OH$ if dissolved in water droplets) to form $(NH_4)_2SO_4$ (see Chapt. 8).

7.  $H_2S$ from anaerobic polluted canals oxidizes to $SO_2$ and $H_2SO_4$.  The $H_2SO_4$
    reacts with the minerals of stone buildings.  Also, industrial $SO_2$
    produces acid attack.  Finally, home coal burning produces soot, and $H_2SO_4$
    content of soot particles is high.

8.  Atmospheric scientists have measured the $SO_2$ and $SO_4$ reaching the land
    surface from rain and dry fallout.  By assuming nearly all this sulfur
    enters streams as $SO_4$, they can account for 80-90% of the S content.
    On the other hand, geologists, working from the S content of rocks and
    annual rates of erosion of the land, get estimates that 50-60% of the
    sulfate in streams comes from rock weathering.

9.  Perhaps a significant percentage of the oxidized sulfur reaching the
    land from the atmosphere is reduced bacterially in the soils to $H_2S$, which
    is re-emitted to the atmosphere and oxidized back to $SO_4$.  Thus it does
    not enter the streams.

10. M.L. Jensen (Science) demonstrated, from studies of sulfur isotopes, that
    the flux of bacteriogenic $H_2S$ from Great Salt Lake as a result of $SO_4$ re-
    duction is a significant factor in influencing atmospheric sulfur in the
    Salt Lake City area.  Thus in this particular instance, bacteriogenic
    recycling was shown to have a larger-than-suspected role.

11. Tidal flats and waterlogged soils, such as rice paddies.

12. No one knows.  It has generally been assumed that pyrite sulfur all goes
    into streams directly as $SO_4$, but the current trend of findings about
    reduced gases and their emissions from soils is that oxidation in soils
    is much less complete than formerly believed.

13. $SO_4$ from gypsum deposits averages about $\pm 18\permil$ (on a global scale) $\delta^{34}S$,
    whereas that from pyrite averages about $-12\permil$.  Thus a 50-50 mixture of
    $SO_4^=$ from these two sources would yield a stream sulfate $\delta^{34}S$ of about $+3\permil$.

14. Because it has been found experimentally that precipitation of gypsum from
    seawater takes place with negligible 'fractionation', i.e., $\delta^{34}S$ of the
    gypsum sulfur is the same as that of the seawater from which it came.

15. No.  Addition and subtraction of gypsum sulfur and pyrite sulfur from the
    oceans would have been constant, so (assuming bacterial fractionation of
    ocean sulfate to make pyrite was constant), no isotopic changes in the
    ocean would have resulted.

16. For the most recent 100 million years, $\delta^{34}S$ of gypsum deposits has remained
    nearly constant, implying a steady state.  But $\delta^{34}S$ dtopped as low as $\pm 10\permil$
    during the Permian Period, indicating major transfer of 'light' sulfur from

the pyrite reservoir to the gypsum reservoir -- perhaps as much as 0.6 x $10^{20}$ moles S.

17. To transfer 0.6 x $10^{20}$ moles S from $FeS_2$ reservoir to the $CaSO_4$ reservoir, 0.6 x $10^{20}$ moles of Ca **are** required (presumably from limestones, $CaCO_3$) and 1.2 x $10^{20}$ moles $O_2$, presumably from the atmosphere.

18. Yes. Today total atmospheric $O_2$ is 0.38 x $10^{20}$ moles. So the question is how to supply 3 times that much to $CaSO_4$ reservoir without removing all atmospheric $O_2$. The continuance of varied life through Permian time (although there apparently were accelerated changes) shows that $O_2$ was never lowered to below-survival levels for most organisms.

19. Perhaps transfer of Ca from $CaCO_3$ reservoir permitted the released $CO_2$ to be changed to organic matter, increasing the rate of burial of $CH_2O$. If so, the excess buried $CH_2O$ would cause addition of $O_2$ to the atmosphere ($CO_2 + H_2O = CH_2O + O_2$).

20. Presumably the ocean. The addition of $H_2SO_4$ or $XSO_4$ to the oceans is minuscule compared to the $SO_4$ already there.

21. In local urban buildups, and in small fresh water lakes. The lakes can become so acid that their biota is altered drastically.


Chapter 8

1. From Fig.30 (carbon cycle), 2500 x $10^{12}$ moles of C are fixed yearly by photosynthesis. Thus the moles N required are 2500 x $10^{12}$ moles C/yr x

   $\frac{N}{C} = \frac{1}{100} = 25$ x $10^{12}$ moles N taken up by land plants each year.

2. It appears that most of the C fixed by photosynthesis is released to the atmosphere by decay and oxidation, and is not recycled in the soil. On the other hand N, after fixation in plants, is largely available again to them in the soil.

3. If 25 x $10^{12}$ moles/yr N are fixed in plants each year, and only 3.1 x $10^{12}$ moles N are 'fixed' from atmospheric $N_2$, then 3.1 x $10^{12}$/25 x $10^{12}$ x 100 = 12% N is not recycled (lost from the system).

4. Yes. If the C/N atomic ratio in living plants is 100/1, and that in decomposed plants is 14/1, then _if_ N remains in the soil, while C goes into the atmosphere as $CO_2$, there is a selective loss of 86 atoms of C or 86% of the C in the living plants, and only 14% of the C remains in the decomposed organic soil material. Thus it seems that N is largely fixed by plants and reused as a nutrient in the soil.

5. Probably not. Much of the N in fertilizer is used in growing high N plants, like alfalfa and soybeans (soybean production = wheat production). These plants are used for animal food. Thus the C/N ratio of the plants

grown using fertilizers is much lower than that used in growing trees, so the living terrestrial biomass is not increased proportionately to the fertilizer N added.

6. $0.71 \times 10^{12}$ moles N from bacterial fixation; $0.57 \times 10^{12}$ moles N by streams; $0.35 \times 10^{12}$ moles N from atmospheric processes (lightning); $1.95 \times 10^{12}$ moles N as $(NH_4)_2SO_4$ and $NH_4NO_3$ -- a total of $3.58 \times 10^{12}$ moles N.

7. N fixed in living organisms = $2500 \times 10^{12}$ moles C/yr $\times \frac{(N)}{(C)} \frac{1}{6} = 417 \times 10^{12}$ moles N taken up by living oceanic biomass. Total entering ocean = $3.58 \times 10^{12}$ moles N. Percent N lost from surface system = $3.58 \times 10^{12}/417 \times 10^{12} = 0.85\%$/yr.

8. Estimates of loss = $2.9 \times 10^{12}$ moles $NH_3$/yr organic decay to atmosphere; $0.3 \times 10^{12}$ moles $N_2O$ to atmosphere; $0.14 \times 10^{12}$ moles N buried in organic materials = $3.34 \times 10^{12}$ moles/yr. A reasonable check on the $3.58 \times 10^{12}$ moles N entering ocean, indicating essentially steady state, with 99% of N reused in surface system of living biomass. (Note: $NH_4^+$ on clays ignored)

9. Much of it (especially in developed countries) goes into sewage and into streams. There it helps increase the terrestrial biomass. The flux of N to the oceans as $NO_3^-$ in rivers presumably also is increased, helping to increase the oceanic living biomass.

10. Residence times are much less than a year, indicating that removal processes are keeping up with additions. Otherwise atmospheric levels would have increased many-fold over the past 10-20 years.

11. No, even if real, the addition is minuscule compared to the total reservoir. In fact, if real, it suggests that some of the N 'fixed' by industry and fossil fuel burning is being 'unfixed', and helping to keep things in balance.

12. No. The total reservoir of $91 \times 10^{12}$ moles is fairly well known. If the residence time were 3 weeks (0.06 years), the annual flux, calculated from $91 \times 10^{12}$/annual flux = 0.06 years, would be $15.2 \times 10^{14}$ moles $N_2O$/yr, far more N than is fixed each year by all processes, natural and artificial.

13. It can be speculated that N fixed by industry and automobiles in the NE U.S. adds sufficiently to the surface waters of the N Atlantic to compensate for the 'normal' N deficit. (Total $N_2$ fixed in oceans today estimated at $3.58 \times 10^{12}$ moles/yr. If N Atlantic is roughly 5%, then total fixed there is $0.2 \times 10^{12}$ moles. If U.S. fixes 30% of all $N_2$ by autos, then it fixes roughly $0.4 \times 10^{12}$ moles/yr NO $\times 0.30 = 0.12 \times 10^{12}$ moles/yr by automobiles (see Fig. 32). If 10% of this ($0.01 \times 10^{12}$ or so) goes out over the N Atlantic and is rained out, the contribution of automobiles would be $0.01 \times 10^{12}/0.2 \times 10^{12} \times 100 = 5\%$ of fixed N going to N Atlantic;

which might put N and P on an equal nutrient basis. This calculation given to show how desperately detailed information is required.

14. $NO_2$ and the final oxidation product $HNO_3$. They are usually listed as $NO_X$ gases. They are toxic and especially noted as eye irritants.

15. Yes. High $NO_3^-$ levels are toxic to infants. Adults rarely are affected seriously. Nitrate in drinking water or food is converted to nitrite in the intestine by bacteria, resulting in illness. High nitrate in food results from heavy use of N fertilizers. (Recent work suggests that high $NO_3^-$ also may be carcinogenic.)

16. $NH_3$ goes into the atmosphere as a product of bacterial decomposition of organic materials. It is highly soluble in water, so in cloud droplets or in rain the raection takes place:

$$NH_3 + H_2O = NH_4OH.$$

The oxidized N gases ($NO_X$) eventually oxidize to $NO_2$, which reacts with water to give $HNO_3$ and NO; $3NO_2 + H_2O = 2HNO_3 + NO$. The $HNO_3$ in cloud droplets or rain reacts with $NH_4OH$ to give $NH_4NO_3$; $NH_4(OH) + HNO_3 = NH_4NO_3 + H_2O$. Also, $NH_4OH$ reacts with $H_2SO_4$ in the atmosphere to give $(NH_4)_2SO_4$; $2NH_4(OH) + H_2SO_4 = (NH_4)_2SO_4 + 2H_2O$.

17. $2.51 \times 10^{12}$ moles N $\times \dfrac{(C)}{(N)} \dfrac{100}{1} = 251 \times 10^{12}$ moles/yr.

18. $251 \times 10^{12}$ moles $\times$ 30g/mole = $7530 \times 10^{12}$ g organic material.

19. $7530 \times 10^{12}$ g/yr organic/density (=1) = $7530 \times 10^{12}$ $cm^3$ organic.

$7530 \times 10^{12}$ $cm^3$ organic/$2600 \times 10^{12}$ $cm^2$ = 2.9 cm/yr organic material accumulated each year. The calculation tends to suggest that a significant fraction of excess fertilizer N could be removed by eutrophication of fresh water lakes.

Chapter 9

1. Phosphorus does not cycle through the atmosphere. The gas phase, $PH_3$, analogous to $CH_4$ and $NH_3$, is an extremely reactive reducing compound, and even if formed in soils or surface ocean, goes directly to phosphate ($PO_4^{3-}$). Also, most natural phosphate compounds have low solubilities in water.

2. Apatite, $Ca_5(PO_4)_3F$ (fluorapatite) or $Ca_5(PO_4)_3OH$ (hydroxyapatite). F may substitute in any proportion for OH; increasing substitution of F markedly decreases solubility.

3. Apparently about 1 ppm F is required. Fresh waters average about 0.1 ppm, but have a range up to many ppm. At 0.1 ppm, human teeth tend to be hydroxyapatite and more susceptible to decay than if converted to fluorapatite. Consequently, many public water supplies have been enriched to about 1 ppm

F, causing a furor with the basic issue of enforced public medication at stake.  If the level in water is raised to several ppm, F becomes toxic and actual tooth and bone damage result.

4.  Apparently not significantly.  Current estimates indicate that the natural flux of dissolved P to the oceans has been increased by less than 10% as a result of man's activities.

5.  The question is only partly resolved, but most soluble P added as fertilizer tends to be fixed in insoluble iron (ferric) and aluminum phosphates in soils.  These are carried down the rivers, when soil is eroded, as solid particles.  Apparently much of the P released to streams, lakes or estuaries from detergents or decomposition of sewage is fixed by increase of the terrestrial biomass.

6.  Yes, especially in lakes, rivers, bays and estuaries.  Growth of water weeds and of blue-green algae has been promoted.  These plants cover the water surfaces, preventing photosynthesis below them.  When they die, sink and decay, they remove $O_2$ from the water, creating conditions for anaerobic decay in bottom waters and sediments.  The result is so-called eutrophication.  The lack of oxygen, and the presence of reduced substances such as $H_2S$, $NH_3$, and $CH_4$, completely changes the biota of the water.

7.  Reduction of use of detergents containing soluble phosphates, removal of phosphates from water in treatment plants, artificial oxygenation (of small lakes), attempts to change lake conditions to promote green algae at the expense of blue-greens.  (Green algae tend to distribute within the water column, rather than forming a surface covering like blue-greens.)

8.  Yes;  many species have the ability to fix N from the atmosphere.

9.  Most calculations indicate that deep ocean waters are saturated with fluorapatite.  Thus the annual additions of P from streams, which eventually, after cycling many times through the oceanic biomass, get into deep water, cause precipitation of fluorapatite into the bottom sediments.  Also, P is incorporated in low but significant concentrations in carbonate minerals and iron oxides.  P is also removed by burial of organic matter.

10.  As indicated before, it is assumed (with little real evidence) that they are incorporated in marine sediments.  However, at the other end of the cycle, when marine sediments are uplifted and exposed to erosion, the P is found in other minerals (apatite, carbonates, etc.).  The details of the fate of the Fe and Al phosphates are unknown.

11.  Probably on the oceanic biota.  Climatic change might affect the rate of upwelling of deep ocean water.  Also, a change in climate would cause changes in oceanic temperatures, changing the solubility of calcium phosphate.  As a result the concentration of P in deep waters might change, as well as its rate of supply to surface waters.

12.  Part of the difficulty is economic;  i.e., what content of P is required to make mining profitable.  Another problem is in estimating the results of further exploration for P deposits.

Chapter 10

1. $2 \times 10^{-3}$ µg/m$^3$/100 µg/m$^3$ $\times$ 193.9 $\times 10^{12}$ g/yr $= 4 \times 10^9$ g/yr.

2. $(0.1 \text{ ng/m}^3)$ $(510 \times 10^{12} \text{m}^2 \times 5 \times 10^3 \text{m})$ $(40) = 1 \times 10^{10}$ g.

3. $\dfrac{4 \times 10^9}{10 \times 10^9} \times 100 = 40\%.$

4. As, Se, Cd, Pb (owing to tetraethyl lead).

5. Review section on effect of trace elements on man.

6. Hg, Se, Cd, As, Pb; they are toxic, found in low concentrations in natural waters, and any addition by man's activities could raise them to toxic levels.

7. (a) $\dfrac{330 \times 10^{15} \text{g}}{200.6 \text{g/mole}} = 1.6 \times 10^{15}$ moles.

   (b) $\dfrac{330 \times 10^{15} \text{g}}{26,000 \times 10^{20} \text{g}} = 1 \times 10^{-7}$ g/g $= 0.1$ ppm.

8. $\dfrac{0.07 \times 10^{-9} \text{g/g} \times 0.32 \times 10^{20} \text{g}}{50 \times 10^8 \text{g}} = \dfrac{22.4 \times 10^8}{50 \times 10^8} = 45\%.$

9. Hg compounds have relatively high vapor pressures compared to Pb compounds.

10. Hg minerals are less soluble than Pb minerals.

11. $.02 \times 10^{-6}$ g/g $\times$ 1000g/day $= 2 \times 10^{-5}$ g/day $= 20$ µg/day.

    Lead ingestion is 0.3 mg/day or 15 times more than Hg.

12. Probably in the oceans. Not likely to be a future peril because on a global basis, organisms do not appear to accumulate much of the Hg. On a local basis there could be more 'Minimatas'.

13. Shark, because it is at the top of the oceanic food chain and feeds on fish high in Hg; reef grazers eat organisms low in the food chain and low in Hg.

14. $\dfrac{30 \times 10^{10} \text{g/yr}}{43 \times 10^{10} \text{g/yr}} \times 100 = 70\%.$

15. Because of inability to experiment on human subjects to determine acute and chronic levels of toxicity, because of long time periods that can elapse before effects of low dosage become apparent, and because of interactions between trace elements.

## Chapter 11

1. Examples are : pro - in emerging nations DDT controls pests that carry disease and destroy crops; con - DDT widely distributed in the environment accumulates in organisms at concentrations shown to be toxic.

2. These birds are at the top of the food chain and their food represents several preceding concentrations of DDT.

3. No. There are $0.38 \times 10^{20}$ moles $O_2$ in the atmosphere. If photosynthesis of marine organisms were cut by 25%, this would amount to an atmospheric $O_2$ drain of at most $625 \times 10^{12}$ moles/yr, or 0.001% per year. It would take nearly 20,000 years at this rate to reduce atmospheric $O_2$ by one-quarter. In this period of time, various feed-back mechanisms discussed in this syllabus would operate to restore balance.

4. $1/2 \times 1/2 \times 8 \times 10^{10} g = 2 \times 10^{10} g.$

5. Atmosphere.

6. One explanation is that some PCB's may originate by degradation of DDT. The patterns of these PCB's probably differ from commerical brands.

7. In 30 years stream and atmosphere fluxes would be 1/4 of today's fluxes and mass of DDT in oceans would be 1/4 today's mass, therefore,

    $$\text{DDT residence time} = 1 \times 10^{11} g/0.5 \times 10^{10} g/yr = 20 \text{ years,}$$

    equivalent to today's residence time.

8. In the centers of the great oceanic gyres (Fig.18); Sargasso Sea, for example.

9. Light and heavy fractions; the heavy hydrocarbons because they are most persistent.

10. Total petroleum transported in 1975 = $10 \times 10^{14}(1 + 0.04)^5$ = $1.2 \times 10^{15} g \times 0.001$ lost = $1.2 \times 10^{12} g.$

11. $0.25 \times 10^{14} g/yr/5 \times 10^{14} g/yr \times 100 = 5\%.$

12. $0.25 \times 10^{14} g/yr \times 0.005 = 1.25 \times 10^{11} g.$

13. Change climate; perhaps lower temperature of earth owing to absorption of radiant energy from the sun by particulates, resulting in less reflected infra-red energy from earth's surface to heat troposphere.

## Chapter 12

1. The total residence time of P in surface waters = P in reservoir/total influx of P/yr; 2710/1.7 + 58 + 998 = 2.5 years. For the oceanic biomass, the residence time is 138/1040 = 0.132 years.

2. $F_{ij} = k_{ij}m_i$, so $F_{23} = k_{23}M_2$, or $k_{23} = F_{23}/M_2 = 63.5/200,000 = 3.175 \times 10^{-4}$/yr (units of millions of metric tons)

$F_{32} = k_{32}M_3$, or $k_{32} = 63.5/3,000 = 0.021$

3. At the new steady state, $F_{23}$ again equals $F_{32}$. Total mass is constant. So, total mass is 200,000 million metric tons + 3,000 million metric tons = 203,000 million metric tons. Then:

$$F_{23} = 1.58 \times 10^{-4} (203,000 - x)$$

$$F_{32} = (0.021)(x). \text{ Since } F_{23} = F_{32}, (1.58 \times 10^{-4})(203,000 - x) = (0.021)(x)$$

Solving, x = 1,520 million metric tons (cut in half).

4. Probably either the land biota or the surface ocean. The land reservoir is too large to be changed much for a long time; the oceanic biota is presumably fixed by N supply; the deep ocean and the sediments reservoirs are also too big to be affected.

5. Residence time of Pb in months = total Pb in µg in blood and soft tissue/ amount of Pb taken in or lost per month.

Total Pb = 3 x 50 x 30 = 4500 µg.

6. More; it is not as soluble as Pb in urine.